MODERN ORIELS

ON ROOFS & FACADES
PLANNING & DESIGN

KLAUS PRACHT

VAN NOSTRAND REINHOLD COMPANY
NEW YORK   CINCINNATI   TORONTO   LONDON   MELBOURNE

Copyright © 1980 by Deutsche Verlags-
Anstalt GmbH, Stuttgart
English translation copyright © 1984 by Van
Nostrand Reinhold
Library of Congress Catalog Card Number
83-10358
ISBN 0-442-27286-3

Printed in the United States

Published by Van Nostrand Reinhold Company Inc.
135 West 50th Street
New York, New York 10020

Van Nostrand Reinhold Company Limited
Molly Millars Lane
Wokingham, Berkshire RG11 2PY, England

Van Nostrand Reinhold
480 La Trobe Street
Melbourne, Victoria 3000, Australia

Macmillan of Canada
Division of Gage Publishing Limited
164 Commander Boulevard
Agincourt, Ontario M1S 3C7, Canada

16 15 14 13 12 11 10 9 8 7 6 5 4 3 2 1

**Library of Congress Cataloging in
Publication Data**
Pracht, Klaus.
  Modern oriels on roofs and facades: planning
and design.

  Translation of: Moderne Erker an Fassade
und Dach in Planung und Gestaltung.
  Includes index.
  1. Building—Details. 2. Oriels. I. Title.
TH2025.P7213 1984   721'.84   83-10358
ISBN 0-442-27286-3

CONTENTS

ACKNOWLEDGMENTS

I would like to thank everyone who helped with this book: the builders and the architects; the various firms and contractors; and the photographers, Dieter Blase in particular. They have designed, planned, financed, realized, and documented oriels, and without them this book would not have been possible.

Particular thanks go to my family, especially Thomas, whose interest in the theme of historical and contemporary oriels inspired me, and whose help in finding examples, taking snapshots, and working on the layout has been invaluable.

My selection of examples was based on the need for quality and the broad interpretation of the theme. Space limitations made it necessary to omit photos showing overall views of buildings with oriels or of roof constructions with oriels. Regrettably, the numerous examples of fine architecture included in such photos would have led us away from our theme and exceeded our limits. On this matter I ask the reader's understanding.

Specific mention is made in the picture credits of the particular features and contributions.

INTRODUCTION

The Theme

Oriels represent attractive architectural elements in historical and in modern architecture. Their efficiency in terms of construction, technique, and craftsmanship is widely recognized.

Oriels have proved themselves on functional and constructional grounds. They are satisfying media for illumination and sunlight as well as for the structural and functional qualities of the interior spaces. Oriels for facade and roof can be used to determine and ornament the architecture of houses as well as of street fronts and plaza facades.

Oriels are attractive because of the abundant variations they permit in construction, insertion of material, realization, and coloring.

Oriels and attics are suitable for every landscape and every house, whatever the purpose or architectural style. They enrich the appearance and increase the practicality of all buildings, whether single houses or apartments, whether one or two stories, whether with flat or inclined roof.

The most varied building materials are suitable for the construction of oriels. They can be realized in the same material as the rest of the building, or they can be emphatically different in form and color.

Constructing oriels is a rewarding problem, but not a light one. New contributions must not be inferior to the older examples in quality. If they are marked by good technical solutions and are consistently and expressively modern, they will also be formally convincing. Old and new solutions can be juxtaposed without the reader (and potential builder) of the proposal having to mourn for the "good old days." Such is not the case with all architectural elements. Doors, for example, rarely sustain the test of historical comparison as well as oriels and attics do.

The examples of oriels shown in this book constitute a representative cross-section. Out of a collection of more than 1,000 items, more than 500 were selected for inclusion.

The Book

This work is intended to provide stimulus for the planning and construction of high quality oriels and roof attics. The subject is one of great contemporary relevance.

After the long ascendancy of flat edifices and sterile facades, after the hiatus in the search for form and ornamentation, there is today a strong need for variety of form. Oriels should thus be studied with the aim of developing new interpretations of old themes.

It is astounding that such a fertile architectural element could have lain fallow for so long, especially in view of the fact that it can be extraordinarily enriching for both interior spaces and facades. In the hectic press of Europe's postwar reconstruction, there was no occasion for such distinct details. It was only after the abatement in new construction that people discovered the charm of old buildings and began to restore them on a large scale.

Today, we have new buildings with strong organic unity, facades characterized by great vivacity of form, and oriels that go far beyond mere utilitarianism. The new oriels are not simply holdovers of old models. New forms are being sought and—as the examples prove—are often found.

Today, architecture is complementary. In the joining together of new and historical buildings and in the building of extensions, modern is juxtaposed with ancient and old is played against new.

The oriels and attics shown in this book are arranged according to their design form, their terminations above and below, and their position in the facade or on the roof.

Edge areas—such as corner windows and balconies, picture windows, and display windows—could no more be omitted from this book than could exterior stairs and porch oriels, since their specific functions and forms relate directly to our theme. The examples of vehicle windshields included in the text are not to be understood merely as amusing diversions, but as hints toward thoughtful solutions for the architecture of tomorrow.

This book should not be treated as a sample book, but should be considered a selection of possibilities for the use of oriels in construction—a selection designed to stimulate new applications. No architectural detail is so small that it can escape being judged on its qualitative merits.

The Reader

Students of architecture and both planning and project architects should find this book of interest.

Architects will find here an overview of the construction methods and variety of oriels and attics. They will of course look for their own new solutions, derived from their own planning work, but they can study the effectiveness of existing oriels and can learn by comparing the examples shown.

Craftsmen specializing in windows, walls, concrete, or glass should find some suggestions for formal solutions relevant to the structures they specialize in. They should thereby gain an understanding that their work makes an active contribution to the overall form of the building. A solidly built oriel is not necessarily a successful one. In addition to technical construction qualities, formal qualities are pertinent.

Builders will find a basic introduction to the architectural theme of oriels and dormers. They should learn to make greater demands on livability and formal quality through variation in interior and exterior shapes.

The book should not, however, be taken as a collection of speci-

Types of Oriel Construction

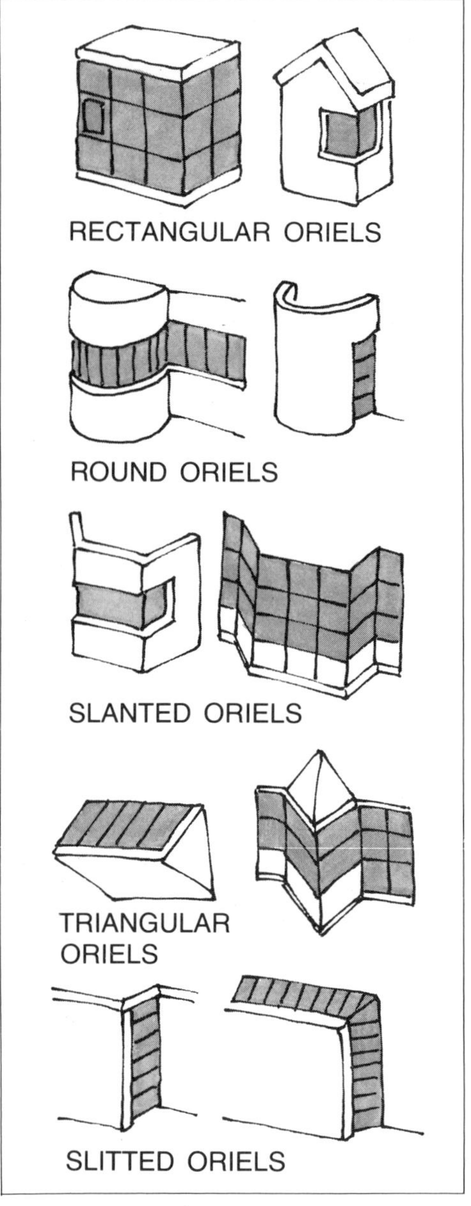

RECTANGULAR ORIELS

ROUND ORIELS

SLANTED ORIELS

TRIANGULAR ORIELS

SLITTED ORIELS

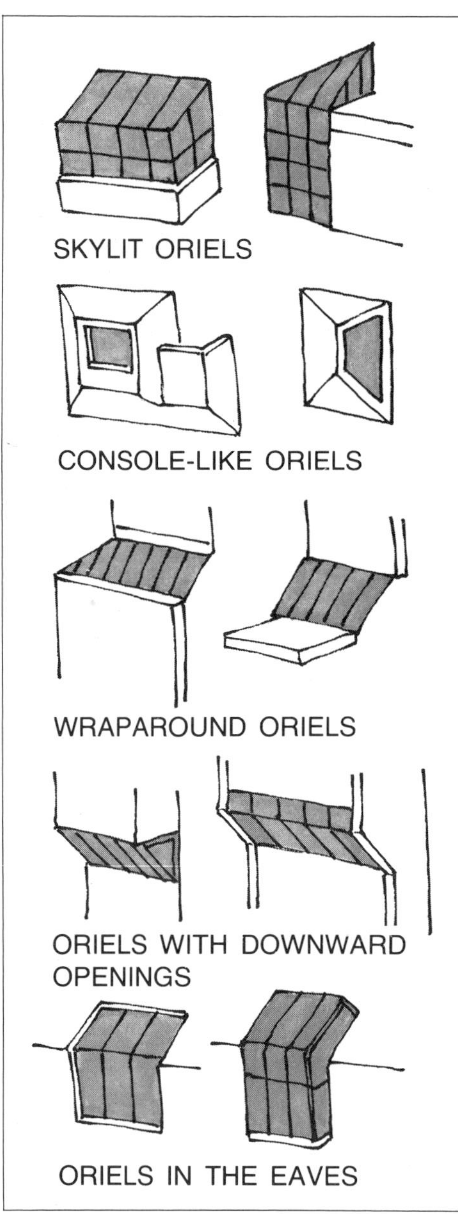

SKYLIT ORIELS

CONSOLE-LIKE ORIELS

WRAPAROUND ORIELS

ORIELS WITH DOWNWARD OPENINGS

ORIELS IN THE EAVES

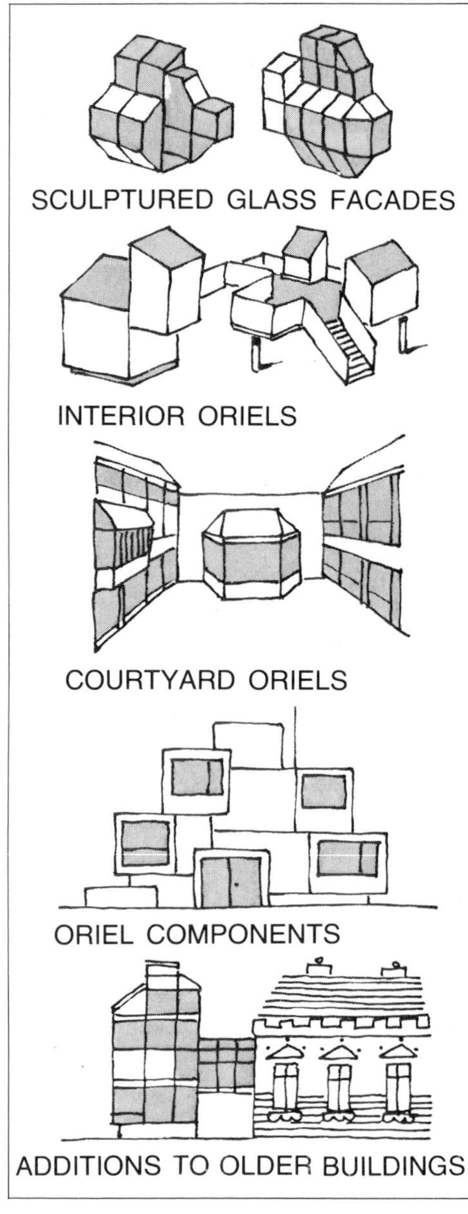

SCULPTURED GLASS FACADES

INTERIOR ORIELS

COURTYARD ORIELS

ORIEL COMPONENTS

ADDITIONS TO OLDER BUILDINGS

(Examples follow p. 16)

Positions

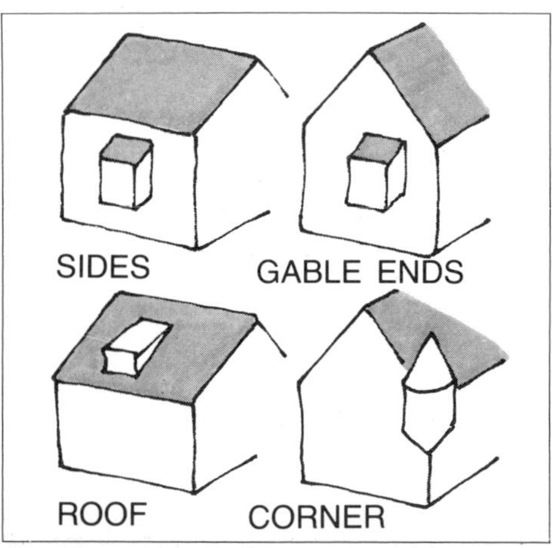

SIDES   GABLE ENDS

ROOF   CORNER

Elevations

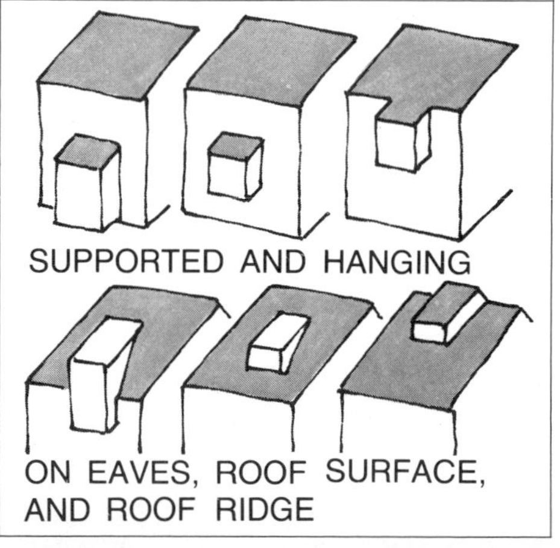

SUPPORTED AND HANGING

ON EAVES, ROOF SURFACE, AND ROOF RIDGE

Glazing

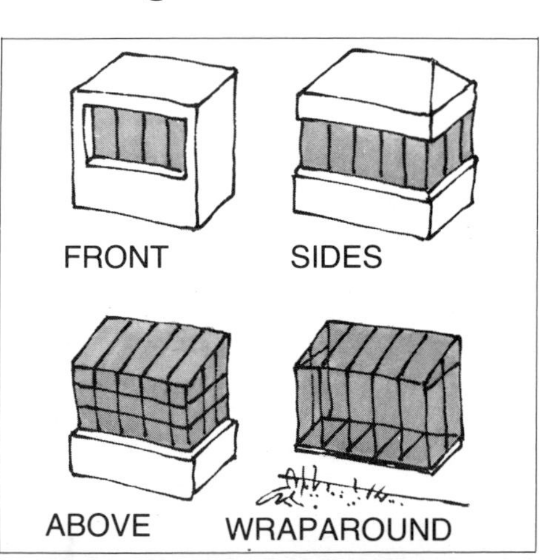

FRONT   SIDES

ABOVE   WRAPAROUND

mens from which solutions can be copied. The builder will come to understand the significance of these particular architectural elements and thereupon evolve his own ideas, which can then be put concretely by the architect and realized by the building company. The best solution to every problem is found through collective effort.

If one bears in mind the specific qualities of proven building materials, the technical realization becomes relatively simple.

The author hopes that, through this book, the planning and construction of oriels will be made easier for all concerned; that the architect will get a feel for the subject, that the builder will accept oriels as a valuable architectural element, and that the craftsman will gain greater understanding of these uncommon structures.

OVERVIEW OF THE THEME

The order of presentation of the oriels in this book is based on the objects themselves and their practical uses, as found in:

Ground floor plans: rectangular, round, triangular.

Angular constructions: straight, sloped, round.

Oriel forms: classic, soft, free.

Extensions of roofs and floors: straight, sloped, round, closed, glassed-in.

Number of glassed-in sides: one, two, three, four, five, or six.

Placement of oriels on the building: on the gable ends or side walls, on the corners or on the roof, on the facade or at ground level.

It is not possible to differentiate oriels entirely from other construction elements. In this book, borderline cases are taken into consideration because, even if they do not belong directly with the subject, they often lead into it.

Corner windows—round, sloped or recessed—permit similar observation opportunities to those provided by oriels. Sloped windows—tilted, flat, bent, or folded—also have much in common with oriels. Balconies are sometimes so built out that they cannot readily be distinguished from oriels.

Oriel Windows

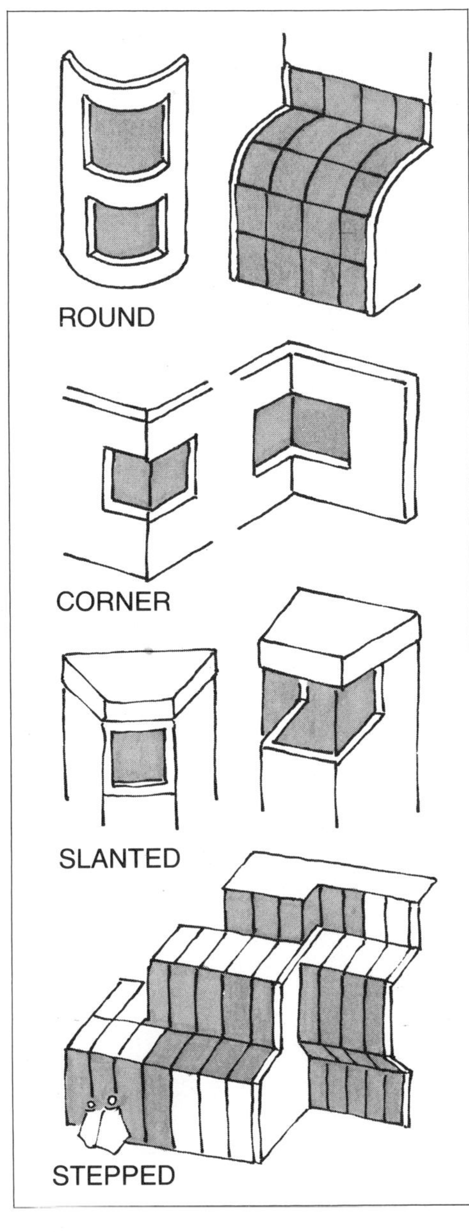

(Examples follow p. 77)

Tilt-Edged Oriels

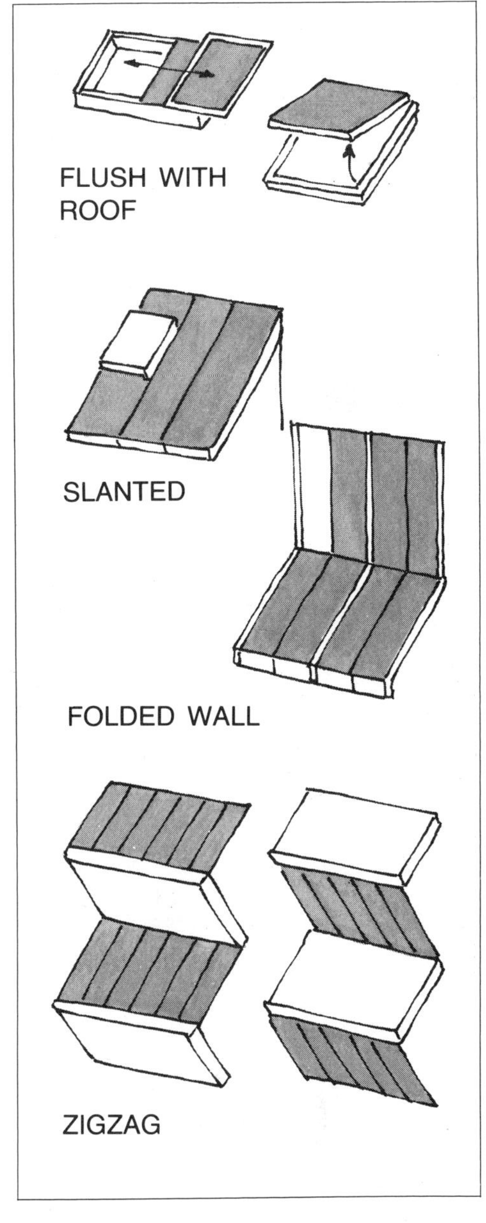

(Examples follow p. 84)

7

Balconies (Examples follow p. 90)

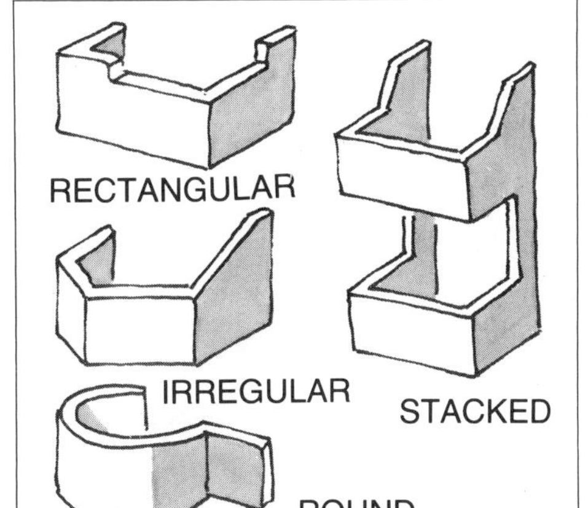

(Examples follow p. 96)

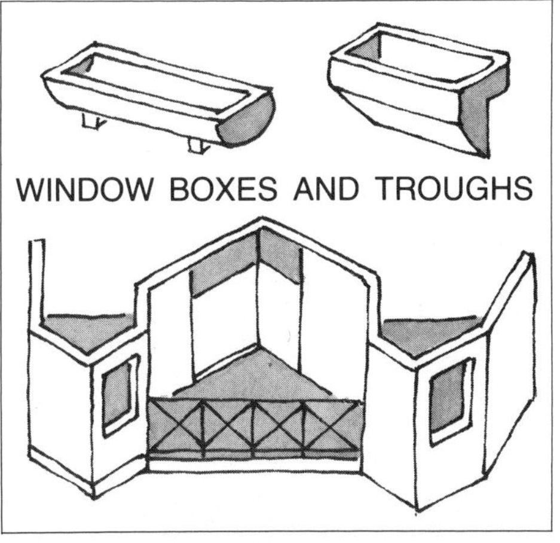

Stairwell Towers

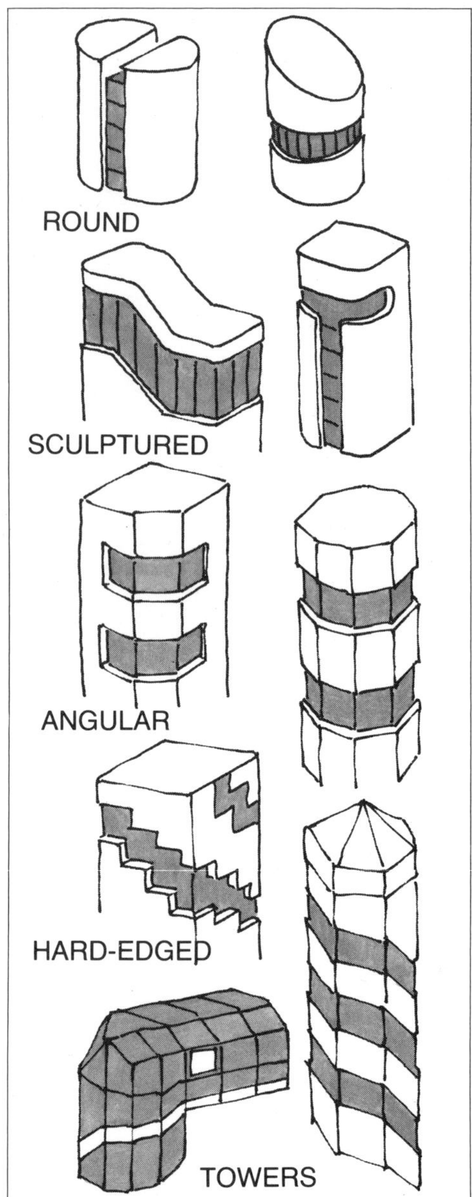

(Examples follow p. 138)

Breezeways

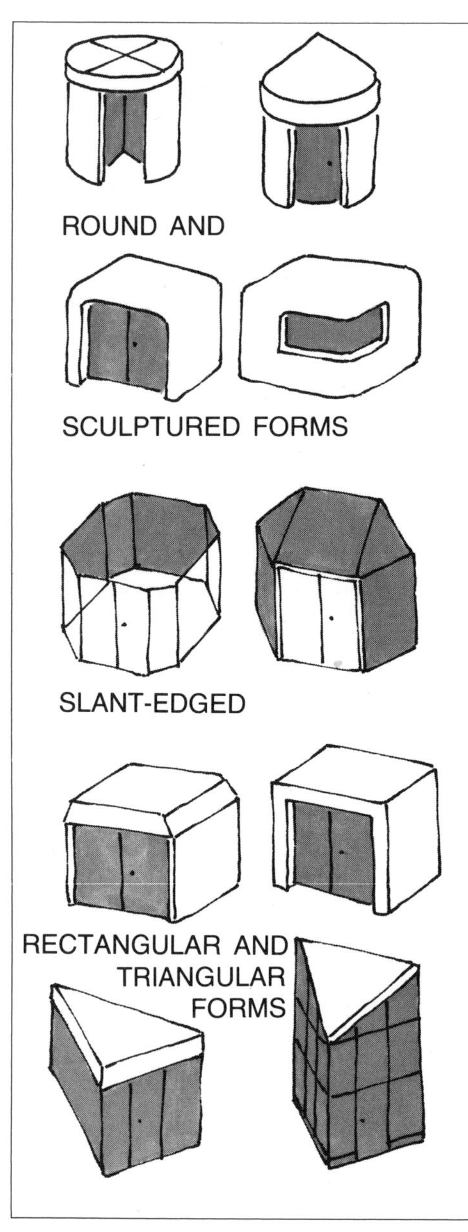

(Examples follow p. 146)

Other notable areas relating to facade extensions include breezeways, show windows, glass cases, porches, glass cupolas, and glassed-in passages. Their particular functions and forms make it possible to talk about them within the framework of a discussion of oriels.

Roof windows (also known as dormer windows) are a particular area within the general theme of oriels. Their importance as decorative roof elements and building accentuations outstrips their functional importance. One can differentiate lean-to- and flat-roofed dormers, gabled and hipped roofs, soft and round oriels, and three- and four-cornered dormers. Skylights, shed roofs, and glass roofs extend construction forms of roof windows almost into the realm of roof oriels.

Show Windows (page 124)

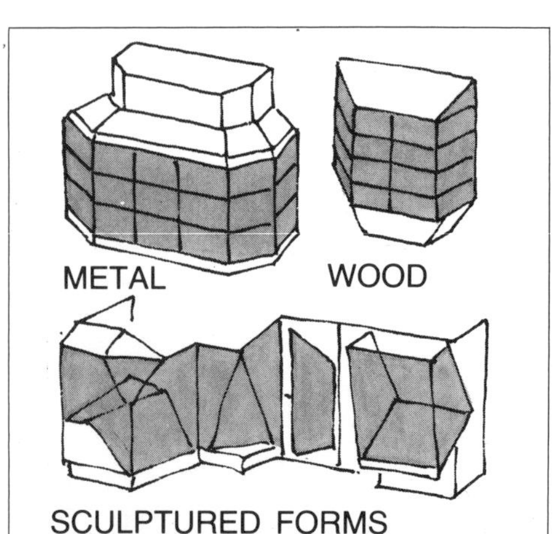

Vitrines (page 130)

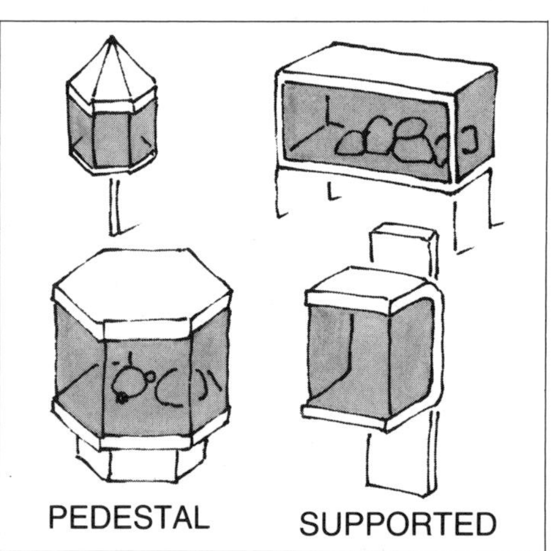

Passages (page 134)

Dormers

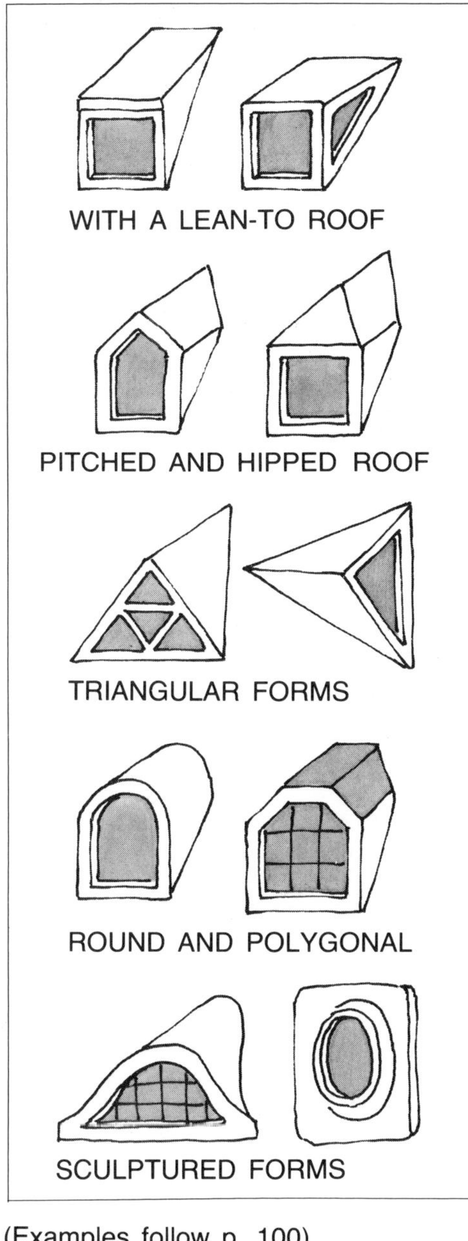

WITH A LEAN-TO ROOF

PITCHED AND HIPPED ROOF

TRIANGULAR FORMS

ROUND AND POLYGONAL

SCULPTURED FORMS

(Examples follow p. 100)

Skylight Constructions

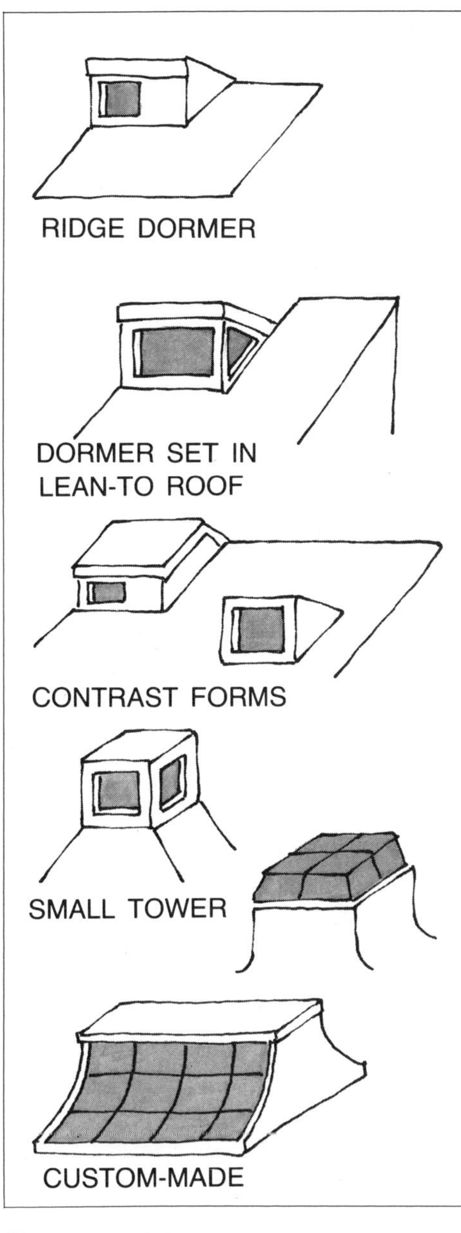

RIDGE DORMER

DORMER SET IN
LEAN-TO ROOF

CONTRAST FORMS

SMALL TOWER

CUSTOM-MADE

(Examples follow p. 108)

Cupolas and Glass Sheds

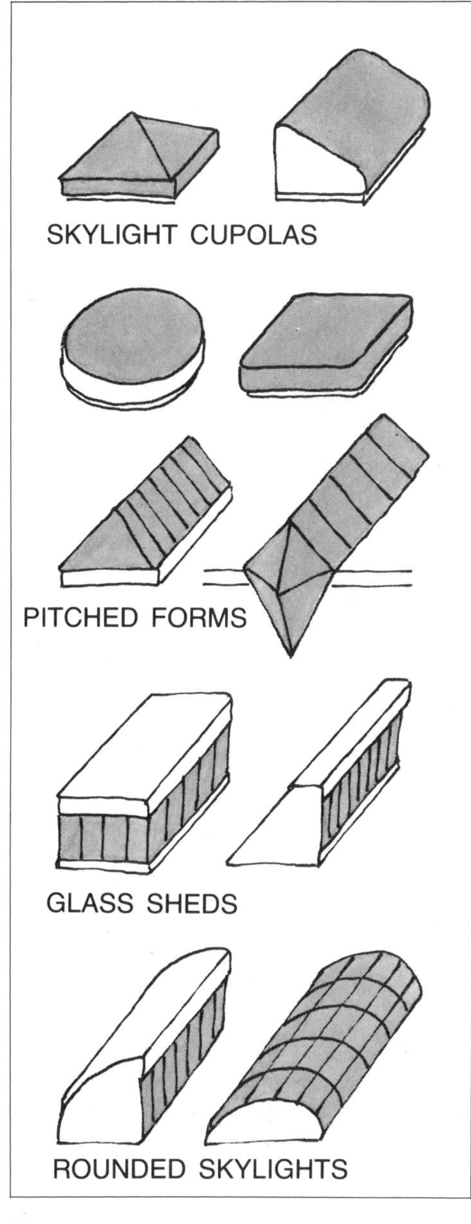

SKYLIGHT CUPOLAS

PITCHED FORMS

GLASS SHEDS

ROUNDED SKYLIGHTS

(Examples follow p. 110)

9

Protruding Roofs (page 122)

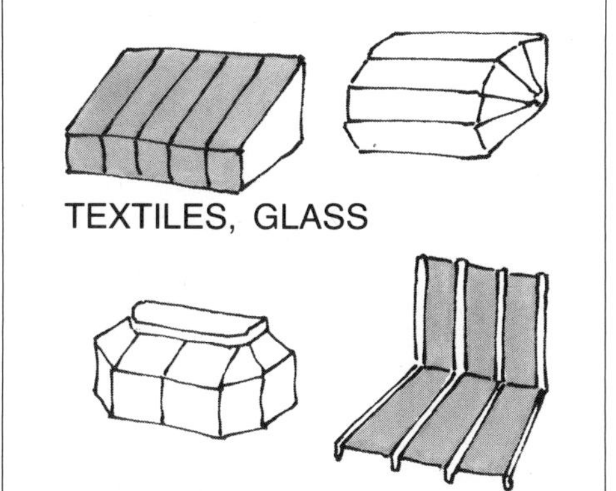

TEXTILES, GLASS

WOOD, METAL, SYNTHETICS

Glass Roofs (page 114)

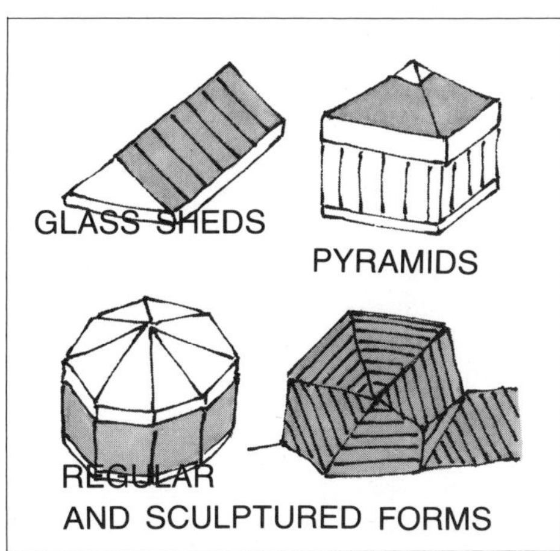

GLASS SHEDS

PYRAMIDS

REGULAR
AND SCULPTURED FORMS

Oriels

Oriels are facade extensions that can be placed on building fronts and corners in a variety of ways. The decision to employ oriels depends on the type of building to be built and the particular functions intended of it.

Oriels are as significant for interior as for exterior architecture. In interior space, they fulfill a variety of functions, the foremost of which is to provide improved visibility. The field of vision from an oriel can be up to 180 degrees, substantially more than that from a standard window. Placing an oriel on a building corner can increase the range of vision to as much as 270 degrees, and a tower-like construction can increase it to 360 degrees. Through the use of oriels, the optimal view out onto streets, squares, courtyards, gardens, or countryside can be achieved wherever appropriate and desirable.

There is a long tradition of shaping rooms through the use of oriels. Surprisingly enough, however, this architectural feature has not been in frequent use for some time.

Oriel forms, construction types, and sizes are as various as their functions. Some oriels begin above table level and some below sitting level, each according to its particular purpose. In general, one can walk into oriels of ceiling height. In earlier times, the addition of stair landings was very popular, and even today, architects strongly favor area differentiation through the use of various levels. Unfortunately, oriels of this type are not easy to realize.

Through placement and arrangement of the oriel and choice of the glass to be used, the amount of light and the quality of illumination can be carefully controlled. Light can be made to fall from above, from the side, or even to rise from below.

Ventilation is not always a feature of oriels: the glazing is often permanent, necessitating that the glass panes be cleanable from outside. When an air vent is considered desirable, it still need not be in the window itself. Slits, grills, mesh guards, or ventilators can be utilized.

In exceptional cases, the need for more room can be a reason to use oriels, either extended at floor level outward or raised into the roof area.

Externally, oriels fulfill various purposes, such as the accentuation of or transition between different building elements. This is equally true for individual dwellings and for street-long facades, as well as for the conception of parks and courtyards.

Oriel-like extensions or gabled porches are widely used as entryway accents, wind- or rain-deflection features, or protective shelters against the elements.

Access to buildings, or between different buildings, can be served by the use of oriel-like entryways or trellis extensions on the facade.

Oriels can be arranged on gables, on side walls, on corners, or under eaves, singly or in groups, and can be successively, horizontally, or vertically coupled, or recessed. It is also possible to juxtapose different sizes and types, or to combine dormers and balconies.

Oriels can be at ground level, on the facade, or hung from the eaves; they can be recessed, free-hanging, or structurally supported. Oriels located on the roof are called dormers, and are treated as a separate category in this book.

The realization of oriels is as various as their functions.

Heating requirements must be considered when one is building large oriels. Over the course of this book, it will become evident which is the most appropriate type of heating: radiator, convector, electrical, radiant floor, or radiant ceiling.

Protection against excessive sunlight, particularly when large glass surfaces are involved, is of the utmost importance. Light exposure, intensity, and duration vary according to direction. Awnings and built-in, roll-down blinds are more effective when used on the outside than when used on the inside. Tinted or insulated glass can also help control excessive sunlight.

Protection from visibility to outsiders, particularly in lighted rooms, can best be handled by the use of curtains, drapes, Venetian blinds or roll shades. The larger the oriel, the more difficult it will be to take into account all the different technical aspects.

Oriels are favorite places for growing flowers, house plants, and cacti. It is therefore important to provide places from which flower baskets, shelves for plants, and even flower boxes can be hung. Herb garden windows and conservatories are particular types of oriel constructions.

It is neither possible nor necessary to categorize oriels precisely; borderline constructions must be taken into consideration.

Corner windows, whether rectangular, sloped, or round, allow much better viewing outlets than normal windows, and lead us further in our study of oriels.

Balconies are sometimes so built up and enclosed that they virtually become winter conservatories. In such cases, it is difficult to distinguish them from oriels. Display windows and show windows also seem to fall into the category of oriels.

Finally, we can use the term oriel in connection with buildings whose facades are so built out or built up with stepped-back extensions that it is difficult to distinguish which portions are extended and which recessed.

The advantages of oriels must be functionally and formally so convincing that they compensate for the difficulties entailed in installing them and for any other disadvantages they may pose. Improved visibility, good illumination, effective room delineation, room extension, and differentiation of building features on facades, as well as other possible functional uses of oriels, justify their frequent inclusion.

These are some of the disadvantages, shortcomings, and difficulties which may result from or inhere in the use of oriels:

Extended portions on facades are, obviously, costlier.

"Pulling out" portions of the facade increases the exterior wall surface, which is unfavorable in terms of heat loss and increased exposure to weather conditions.

Dust and dirt collect on extended features of the building, and dirt is particularly noticeable on sloping glass surfaces.

Maintaining and cleaning differentiated exterior walls can be complicated, and, if access is not well thought out technically, they can be downright dangerous.

Extensive use of glass surfaces makes it difficult or even prohibitive to economize through insulation.

The technical planning required to fit out an efficient heating system and to protect from sun and glare can be complex. These crucial factors must be brought to the attention of builders and planners.

Dormers

Roof constructions with windows (oriels on roof surfaces) are designated as dormers. They are also called gabled or garret windows.

Dormers provide a distinctive effect both for the exterior of the building and for its interior. Here are some uses for dormers and some of their varied aspects:

Dormers allow for the illumination and ventilation of attic rooms. They are particularly useful when these functions are not met adequately by windows placed only on the gable ends. In contrast to many oriels placed on the facade or at ground level, dormer windows are not per-manently closed, as they (unlike oriels) must be cleanable from the inside.

Normally, the prospect seen from the attic room is rectangular in format and parallels the roof ridge. Nowadays, it is popular to have side windows as well.

The room extension provided by dormers in attics is vertical, a means of heightening low ceilings.

The room design possibilities are not as striking for dormers as for oriels, at least not as a main point of departure in planning. However, dormers are popular since they contribute, like oriels, to making a room's atmosphere agreeable. In fact, they are even more frequently used than oriels because builders can convincingly argue their value in increasing illumination in attic spaces—an argument that cannot be applied so easily to oriels.

Variation of the roof surface has a considerable effect on a building's outward appearance. This can influence even the appearance of the street facades, rooms, and courtyards.

Dormers can be arranged singly or in groups, coupled or recessed into the roof, on the roof ridge or at eave level.

The location of dormers is variable: they can stand on the roof story or can hang from the roof surface; they can be placed at the joining of two rafters or, in exceptional cases, can be drawn upwards to the roof ridge; or they can rest directly on the outside wall or on an apron. The precise delineation of different dormers is difficult—they are roof elements, just as windows are.

The appearance of a roof window can be varied by placement on a sloped roof, or on a flat roof as a glass cupola, skylight, or glass "roof."

Buildings utilizing dormers range from small constructions to large structures completely made up of dormers and oriels combined. Special dormer variations are used for ventilation and even for exhaust purposes in storage rooms and pantries.

Disadvantages, weaknesses, and difficulties:

Openings in the roof are expensive.

Roof drainage systems will be impaired or even rendered inoperable.

Drainage systems for dormers themselves are often complicated.

Snow build-up can be caused by the presence of a dormer, particularly if a hollow is formed where it joins the roof. Defrosting mechanisms using heated wires or mesh are at best makeshift solutions. Making the points of contact between roof and dormer watertight is always problematical.

The design of dormers is not always satisfactory. For example, in the case of sloping windows placed even with the roof surface, there is often distortion from inside and outside due to their being placed so low on the roof, and this makes for a bad effect.

Advantages:

Good illumination of attic rooms.

Better views than from sloping windows placed even with the roof surface.

Room extension for low-ceilinged attics.

Good design feature for interior space and for roof ridge profile.

There are, then, good reasons for the inclusion of dormers, despite financial considerations and the technical difficulties entailed.

Bird's-Eye Views and Profiles

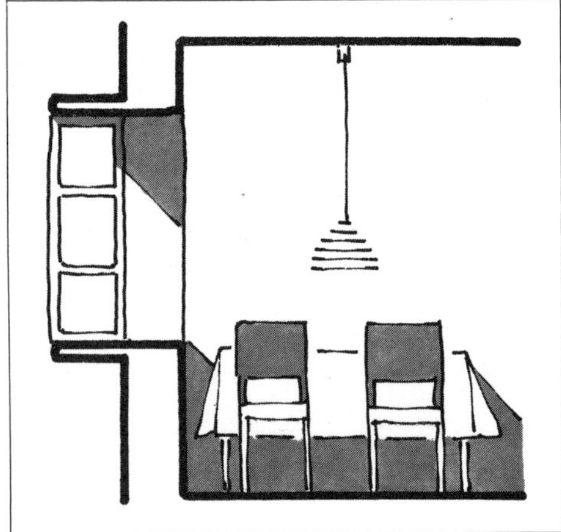

The amount of light that enters a room through an oriel has a greater influence on the room's atmosphere or charm than does the oriel's form or shape. The myriad construction forms available are not widely known and are rarely put to advantageous use, even by experts.

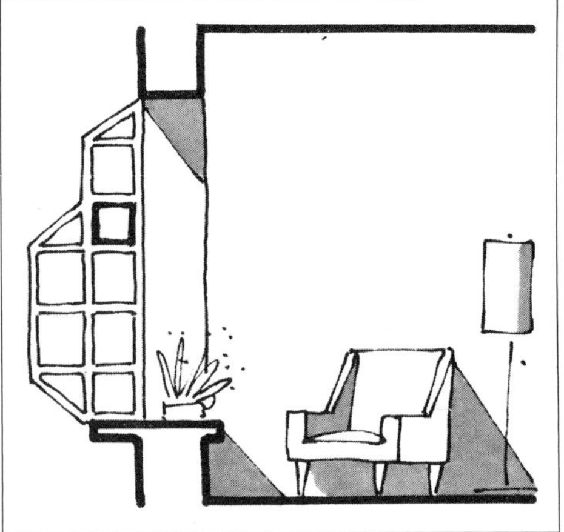

Illumination from the front allows excellent views, which can be optimized by the use of side windows. Illumination exclusively from the sides, from above, or from below makes for interesting and arresting effects, even if the view afforded is unorthodox.

Glassing-in the floor while leaving all other surfaces of the oriel unopened seems absurd. However, one can imagine that the view from such a vantage point down into the street might resemble a painting hung in the oriel itself.

Oriels can be implemented at different levels.

At table level, they allow for the coordination of a table and sitting group. Therefore, they are used mostly for dining rooms or dining nooks, or for alcove rooms close to work areas.

At sitting level, oriels afford better views and, depending on their depth, may actually allow one to sit farther out than the bearing wall. In this form, they serve particularly well as plant corners, usually with a raised sill or window garden.

Oriels that are as high as the room itself can be entered. In older oriels, one often finds a platform or slightly raised landing. Nowadays, unifying floor surfaces are preferred. The examples demonstrate even, raised, classical, and free-form solutions. The variety of forms in oriels is virtually numberless.

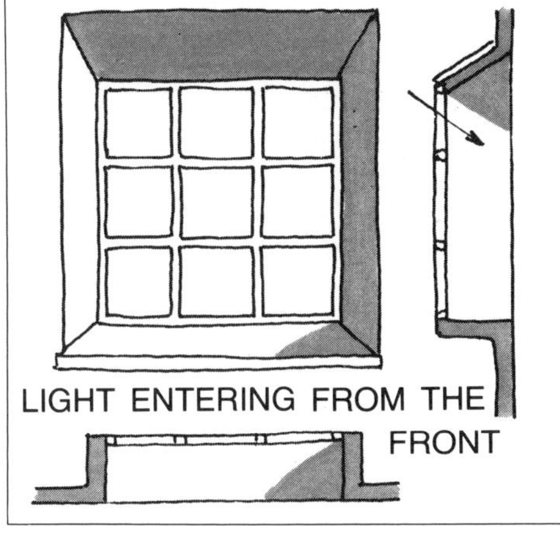

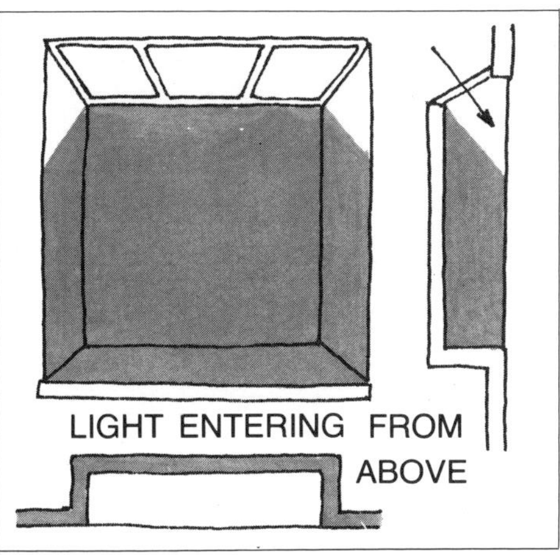

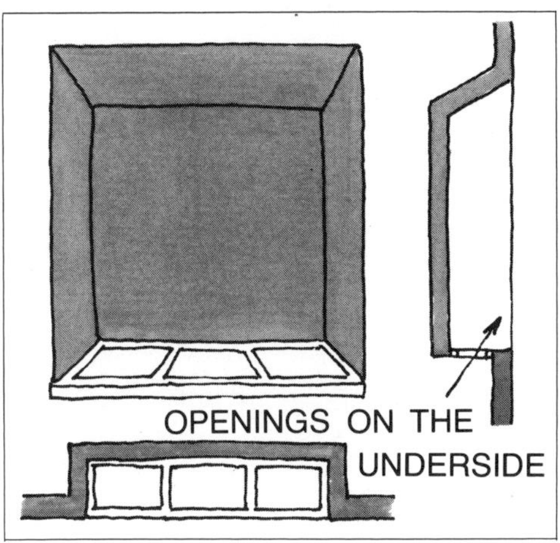

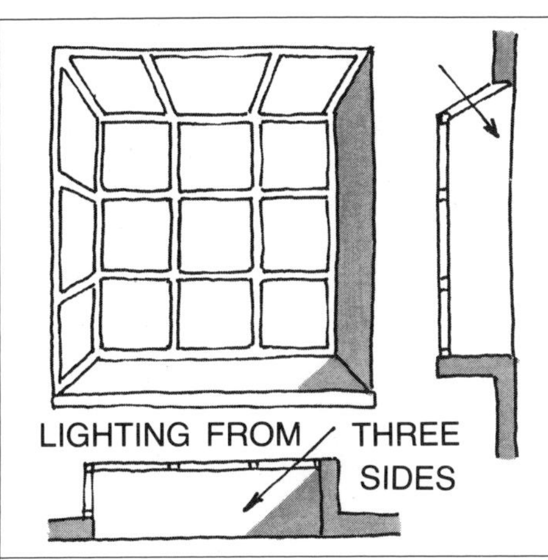

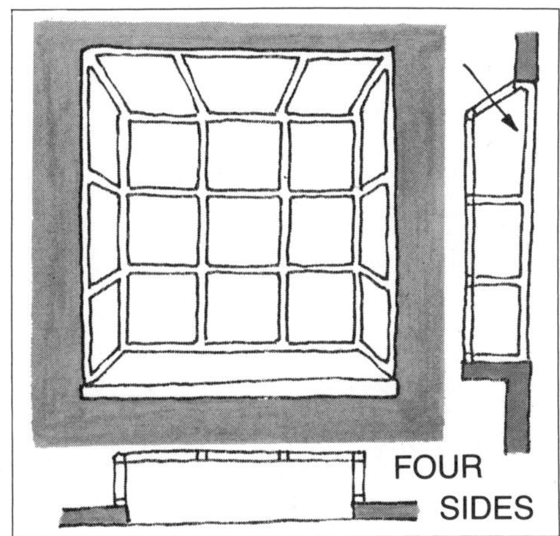

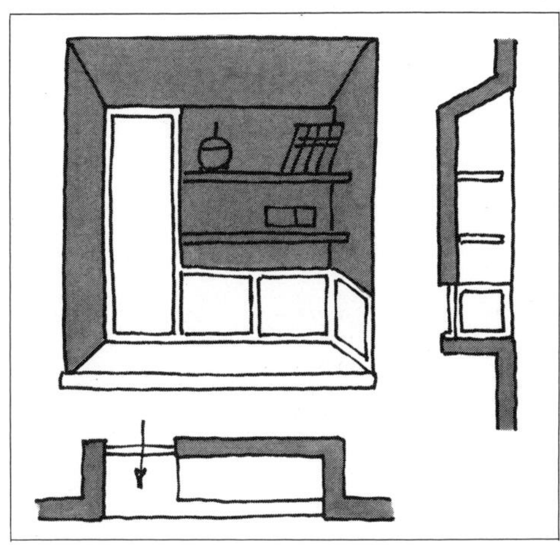

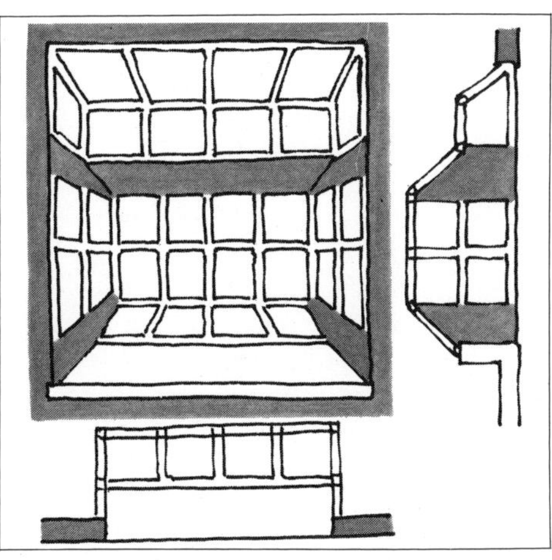

Custom-designed, plastically formed, and partially glassed-in oriels:

The left-hand example shows a well-illuminated working area without glare and with shelves above. The vertical window openings made of glass strips permit an unobstructed view.

Right: Oriel as a free-standing extension on the facade.

Design

The design and coloration of oriels (and dormers) are as multitudinous today as they were in earlier periods. This single architectural element manifests itself in a richness and variety that allow the design originality of architects, contractors, and builders to be clearly read. The extreme variety of design is due to the demands of interior space and to the specific placement of the oriels on the facade.

On the floor plan, one can distinguish rectangular, triangular, sloped, polygonal, semicircular, and circular oriels. The upper extremities use either flat, lean-to, or gabled roofs.

The lower extremities can also be flat, with or without a balcony support, or sloped like a console. The overall design of the facade dictates whether the oriels should be solitary, grouped, in rows, or staggered.

The use of color in oriels remains intensive, although more unified than in the historical examples, which are often extremely colored and sometimes extensively ornamented too.

Coloration can derive from painting, impregnation, anodic oxidation, or from the color of the building material itself. Possible materials include wood, aluminum, steel, concrete, natural stone, and brick. Synthetic materials also play a part.

The glass surfaces are clear, ornamented, or opaque; toned to cut glare; single- or thermal-paned; large or latticed. The shapes of the windows may be quadrilateral, rectangular, polygonal, or round. The number of windows varies; they may be solitary or grouped in patterns.

Glazing is particularly important in oriels because the amount of light entering these facade extensions through the various planes of window openings varies enormously, and these variables powerfully influence the atmosphere of the room.

The lighting effects must be planned with great care; they are at least as important for the interior design as for the exterior architecture of the oriel.

The arrangement of the windows can be variously combined, according to individual taste. Light can enter from the front, from above, from the sides, from all of these directions at once, and even from below. This last is a rarity, however, since one could not put anything down onto the oriel unless the glazing were especially designed to carry weight. (Technically, this is perfectly possible through the use of strong shatterproof glass.)

Design of dormers is as varied as that of oriels. Originally, dormers served chiefly as ornamental elements of roofs because attic rooms were not as large as they are today. In our times, it is the illumination potential of dormers that determines their inclusion in plans. The richness of design possibilities is thus no less important now than before.

A distinction can be made between free-standing and supported dormers, and between those with flat roofs and those with lean-to, gabled, or sloped roofs. Polygonal or rounded dormers are rare, although they are particularly eye-catching and usually conform to the design demands of roof and facade.

Dormers utilizing side windows afford a view that parallels the roof ridge, and usually do so more for the sake of function than of ornamentation.

Planning

Ideally, the planning of oriels should proceed from the interior to the exterior. The requirements of the interior space dictate the functional modes. Producing the resultant exterior design is not, however, an easy task. Often, precisely the opposite process occurs: the architects are so intent on a rich exterior presentation that their thinking goes no further. According to today's architectural precepts, it is certainly preferable to consider the functional needs and formal qualities of the interior space as the point of departure.

The function of these facade and roof extensions is to be extrapolated not only from how they are utilized, but also from their design elements. It is often the design of particular construction elements that makes their use possible or that enhances them.

Recently-constructed buildings feature an astonishing number of oriels and dormers. Even when there is no need for an oriel, one may be painted on the facade. What was prohibited after the war is now prized as "street art."

Newer examples are in no way inferior to earlier ones in quality, persuasiveness, color, design, or variety of materials.

Old oriels and dormers are no longer casually torn down or modernized; instead, they are carefully renovated. The love of the old-fashioned has not gone so far, however, as to lead to the reconstruction of oriels.

Combining new oriels with older buildings is an especially attractive architectural exercise. In doing so, we can be true to the lights of our own times and yet operate in a manner similar to that of older masters and more ancient epochs. For example, during the Renaissance, oriels made of sandstone were artfully added to medieval brick buildings in such a way that, with all the contrasting elements, an overall architectural unity was achieved. Their persuasiveness can be monumental. To cite just two examples, the city halls of Bremen and Lemgo are outstanding.

Materials

In the making of oriels, standard materials such as stone, wood, and metal are used, as well as modern synthetic materials in the form of mouldings and plates, or acrylic "glass."

Wood, either carved or plastered over, is very well suited for the task, since all mouldings that are frequently used in this mode of construction are available to medium and small firms and are thus readily available even in the country. The wood can be protected by impregnation or painting.

Steel, which is plated and then painted, is often chosen for oriels because of its non-massive quality, which offers a slender profile. Galvinization protects against corrosion. Colors are painted on and can be as varied as those used with wood.

Aluminum, in natural or accented tones, protected by the process of anodic oxidation, is especially prized for its precise contours and for its heat retention.

Concrete is used primarily structurally to carry weight, but can also be textured or used in slabs.

Brickwork is more than regional in significance. Bricks are used on oriel ledges, supports, and cornices in such fashion that, with the addition of unique elements, highly individualized structures result.

Synthetics serve primarily in custom constructions. Extending porches or decks built to cut glare can often be recognized as facade extensions, or oriels. In these structures, synthetics are utilized as mouldings, contour elements, slabs, thin skins, or grooved elements, or as solid overhangs.

Single-paned glass is seldom used as a construction element, having been superseded by thermal-paned glass because of the latter's heat retention and reasonable cost.

Acrylic glass is used when unusual forms are required, as in the case of slightly bowed windows or of windows on roofs. Skylights made of plexiglass are a widespread example.

These materials are used singly or in combination with one another—for example, aluminum combined with wood. Steel strengthens aluminum constructions; concrete supplements brick. Synthetics and glass surfaces are made rigid through the addition of different materials.

The choice of materials is crucial in the design and cost estimation of buildings. If the individual characteristics of each material are not considered in the design and engineering, structural damage can ensue. These considerations often totally control the variables of special constructions, and therefore, ultimately, the design. Construction must be understood in this light and kept in hand.

However, architectural design should not be narrowed by such technical considerations to a point at which original solutions cannot be sought. We must pursue the need to use new materials in original ways. Such has always been the case in the past; new stylistic epochs have developed new forms that would have been difficult to attain using older technologies. People have met these challenges by inventing veneering techniques for marble and other materials. Today, the desire for original oriel forms has become so great that we can speak of wholly sculptural structures.

Construction

Oriel and dormer constructions can be complex, but more often they are simply regular window construction features or are elements of large superstructures. Construction techniques are therefore standard, well-known, and—except in special cases where the connection of surfaces is sloped—simple.

The manufacture is done by firms that excel in specialization rather than in magnitude of production. Oriels differ in size, weight, and function, and construction techniques vary accordingly: some oriels are on ground level, some on roof stories, and others hang from facades.

The structures we call oriels are made up of fronts, sides, and upper and lower elements that serve as roofs and floors.

The walls either are unified frameworks with glass surfaces or are divided, with the window sitting on its sill and wall. Glass surfaces are either fixed or moveable, depending on whether they can be cleaned from outside or not.

The floor can be of different heights: table, floor, or seated eye level. Some can be entered, others not. The engineering must therefore take into account these different weight-bearing loads.

The roofs can be enclosed or glazed, flat or sloped. Its weight is borne by the sides. In construction planning, it is necessary to keep an eye on the relative costs of weight-bearing elements, and on whether they must be fixed or moveable.

Cantilevered oriels are usually made of lighter materials like wood, aluminum, steel, or even synthetics. Massive oriels are usually built at ground level, stacked on one another, or supported. The depth of the oriel is the decisive factor in determining whether it should cantilever or be supported.

PARTICULAR FORMS

We take as our model the craftsmanship of old masters who created optimal solutions, balancing technical requirements and aesthetic preferences. When we look at historical examples and compare them with our own time's contributions, we can rest content if the latter bear the comparison well.

It is remarkable to us today that, on the one hand, constructively necessary elements such as onion domes, bow fronts, supports, and buttresses dictated overall appearances, and yet that, on the other hand, these very design elements took into account the taste of the times. Wall plates and bressumers and studs, for example, were so formed or so covered by carving that their constructional function appears underplayed.

Half-timbered oriels on wooden or massive constructions are among the most beautiful historical examples. The carving and painting make them especially admired and renowned. Often the oriels were added later, and yet there is no disruption, even though they stand out clearly from the original construction.

The oriels shown here all have rectangular floor plans. However, their differences in other respects make comparison difficult.

It is of paramount importance in planning to imagine the various interpretations possible in three dimensions.

Flat, stepped, or sloped roofs close off the oriels above.

The oriels stand at ground level or on supports, hang on the facade, or appear stretched between walls.

The materials—brick, limestone, wood, and metal—dictate the character of the oriels, as do their colors and textures.

Visibility is varied by the arrangement of window panes or mullions.

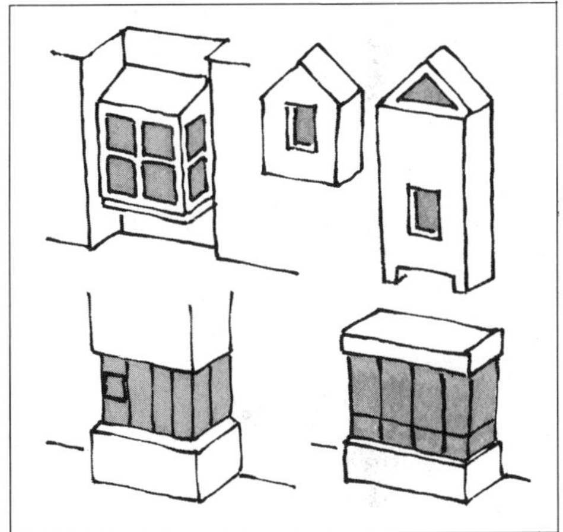

19

Rectangular Oriels

The arrangement of oriels greatly influences the design of a facade. Stiff groupings contrast with free compositions. Here, single elements are collected into complexes; there, large building masses undergo an agreeable differentiation. Alternation of colors and materials improves the composition through extensive articulation.

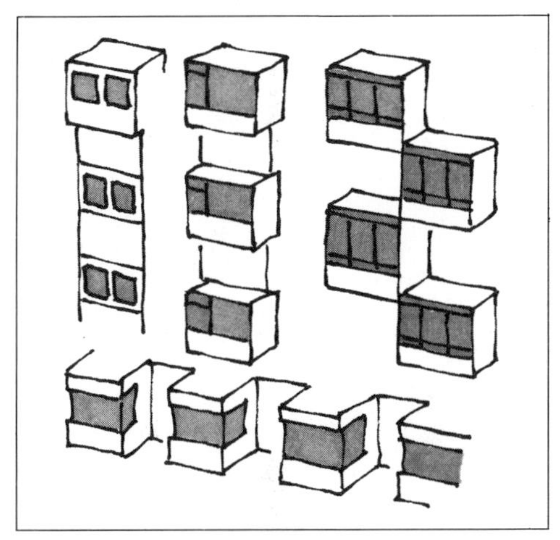

Groups and Rows

Old and new oriels confront one another. They underline how old and, at the same time, how contemporary this theme is, and they show how oriels are successfully interpreted in each new epoch.

The differences are also based on choice of materials. Bricks have served Jugendstil and Bauhaus alike. Today's architecture is noted for huge glass surfaces and the use of synthetic materials.

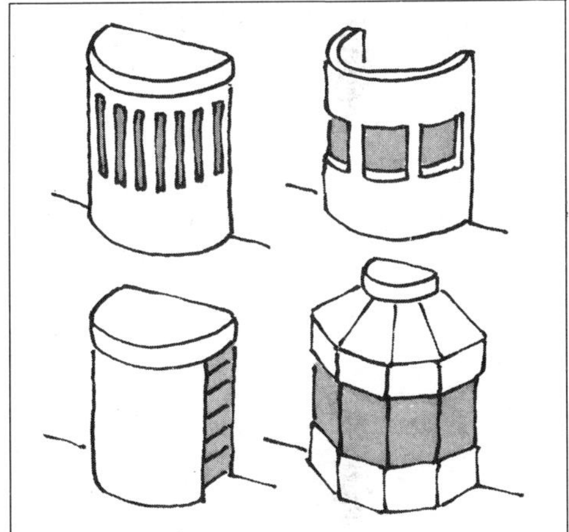

The closely arranged window surfaces of most of the examples shown here are due to the difficulty of glazing round surfaces with flat panes.

An extreme exception, at left, is the round, completely glassed-in oriel.

Half-Round Oriels

The sculptural quality of these tilt-edged oriels is emphasized in the way the rather deeply inset windows are joined into running bands.

This was a technical necessity, due to the fact that exterior roll blinds are hidden under the sun-deflecting overhang.

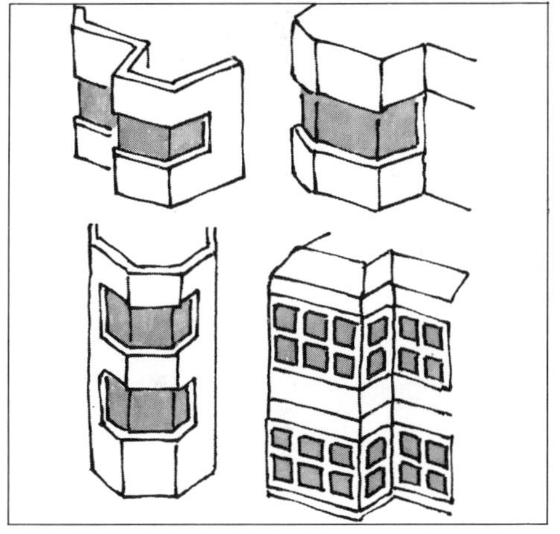

Tilt-Edged Oriels

Above: Smooth concrete slab construction, optically heavy, is contrasted to unmullioned glass surfaces on ground level.

Below: A sharp combination of color and material. White story supports, red brickwork siding, aluminum window frames anodized in green, and blue-gray roof slates make a colorful composition.

An arrangement of oriels, balconies, and a roof garden is covered by pergolas. It shows what is possible when imagination is used, building regulations are stretched to their limits, and officials make their decisions in a positive and creative fashion.

This rather monotonous street scene is pleasantly accented by the division of the windowpanes.

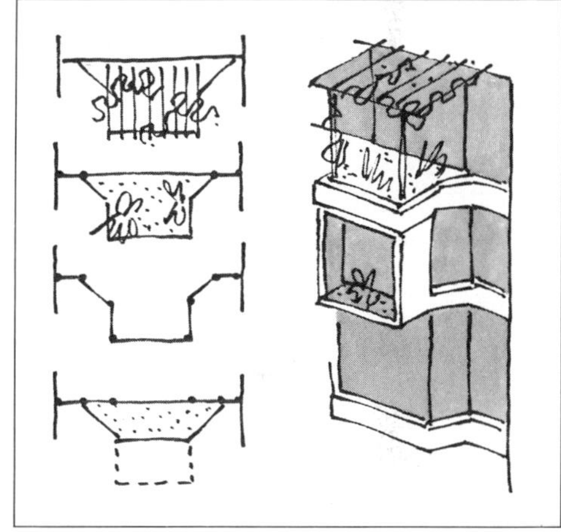

With Plants

Poured concrete supports carry window boxes. The oriel extension hangs by steel bars, which, with the small, glassed French windows, results in an airy effect. Steel cable stretched between supports meets safety regulations without blocking the view. Spotlights illuminate the plants at night, thus optically enlarging the living-room space.

Design and construction, or construction as design, appears to be the theme here.

Supports offer the opportunity of glassing-in areas or of leading the glassed-in areas around supports.

Story ceiling supports and extension supports require slanted window frames.

The materials emphasize contrasts between the architectural elements; for example, the oriel fronts as glass-bricked walls and, at left, the triangular oriels realized in brick.

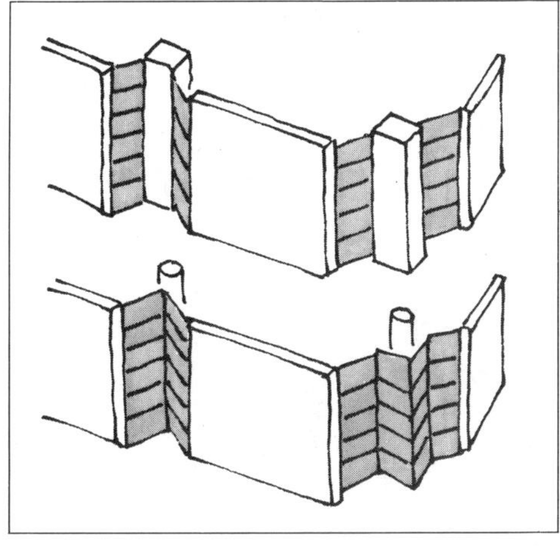

29

Triangular Oriels
With Supports

Triangular oriels differ greatly because of the different treatment of details.

The upper closure is a stepped or sloped roof.

The lower closure is straight or slanted. Illumination comes from the side, through front windows, or through skylights.

Materials: Brick-laid limestone, asbestos-sheeted veneer, wood or metal window frames, simple or thermal panes, clear or tinted.

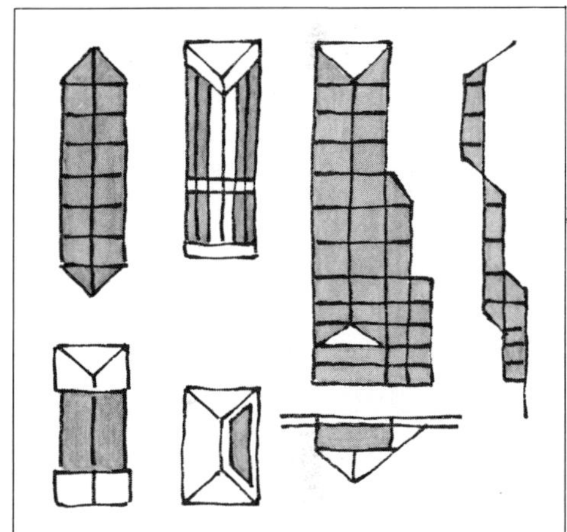

With Lean-To Roofs

Very effective forms are achieved through the use of non-quadrilateral bodies and construction elements that are pushed outwards from the facade.

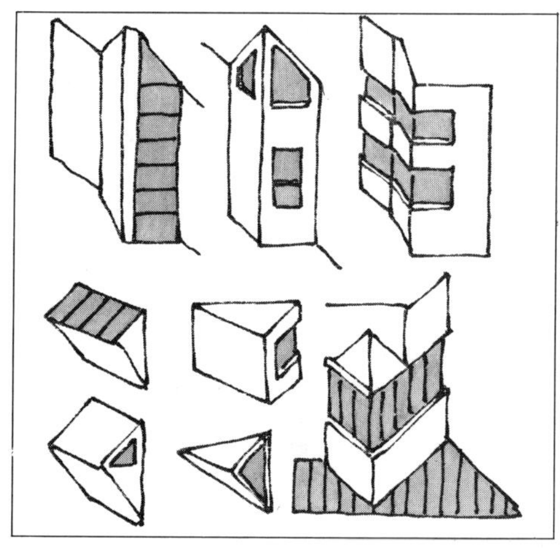

33

Custom-Designed

Facade oriels and roof dormers constructed of brick are among the more unusual building details of previous periods. Such "material-specific" architectural forms are considered exemplary today.

The arrangement of individual elements, the saw-toothed profiles, and the use of supporting brackets or pedestals are typical in their brickwork and characteristic of their times.

35

Slanted glass surfaces produce attractive "oriel sculpture," both in the floor plan and in elevation. The choice of materials can emphasize these effects.

Right: Open, cinder block wall contrasts with the facade made of corrugated, poured concrete component slabs.

Above: Aluminum (anodized black) contours are filled in by reflecting glass surfaces, which can be protected from glare by rolling down yellow-tinted, see-through shades.

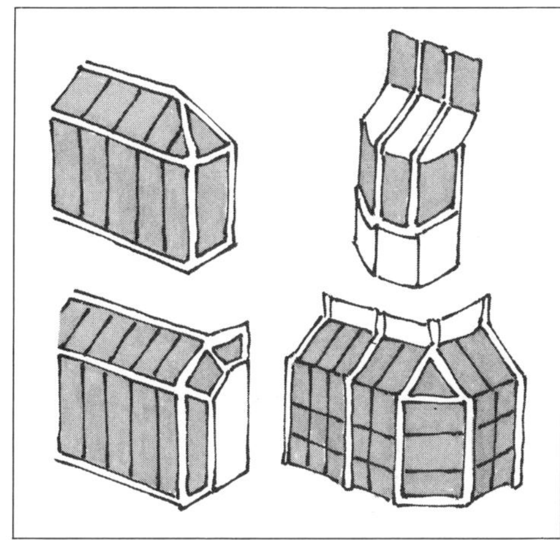

Skylit Oriels

Extended balconies and exaggerated oriels are combined here.

Left: An oriel with balconied loggia on top.

Right: A balcony, extended on all floors from the facade, capped by an oriel construction.

Protruding balconies—above or in front of bay windows equipped with slanting or rounded racks for creepers—again are in the style of alcoves, a play with shapes and materials producing abundant diversity.

Filigree solutions (left); very robust execution (right).

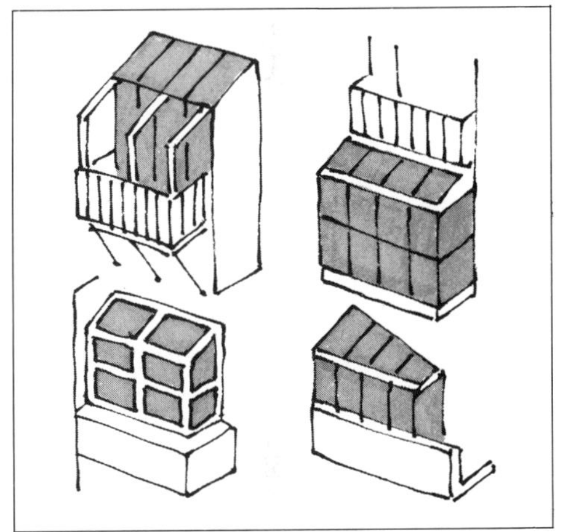

39

With Balconies

Oriels that can be entered, terrace extensions, windows for plants, and winter gardens are shown here. The immense glass surfaces, which lead at left to the garden, combine the green of the garden with the living area, as in the conservatory oriel at right.

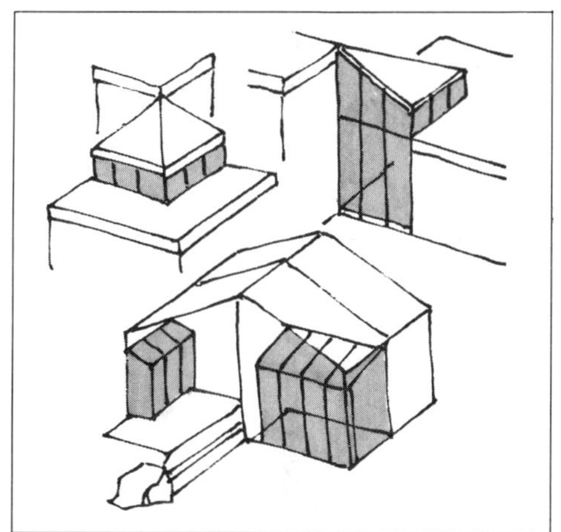

Terrace
Oriels/Conservatories

Oriels such as the ones depicted here, which conclude in tilted surfaces, are designated as console-like.

The heavy, poured concrete extensions actually fulfill a structural function, while the light aluminum reflects a more aesthetic desire to anchor the airy construction to the building.

People spare no effort in exploiting to the full the possibilities of oriels.

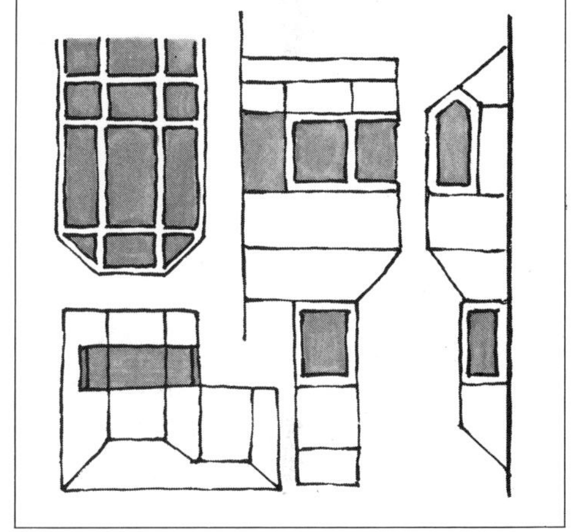

43

Console-Like Oriels

This school library has a well-lit and friendly atmosphere due to the oriels opening outward onto the garden.

These penthouse apartments demonstrate oriel-like studio skylights featuring large glass surfaces. Design and construction successfully achieve their goal of producing something out of the ordinary.

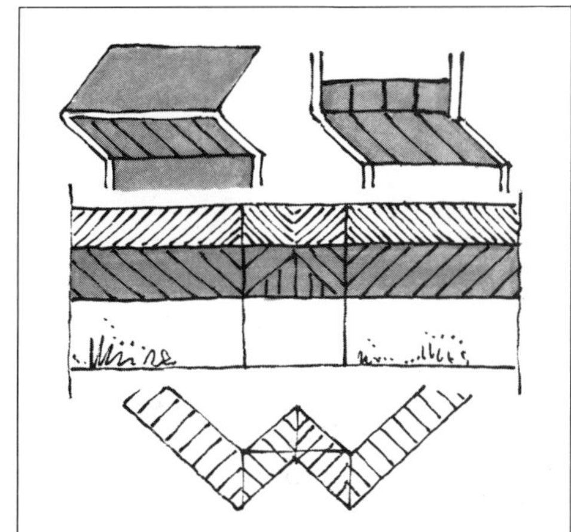

45

Lit From Below

The skylights, running from the extended roof line to rejoin the perpendicular end walls, form an oriel-like window construction.

The horizontal band of glassed-in surface connects the oriel extensions and the walls of the building. Vertical fields of windows step upwards to the lean-to roof ridge. There is hardly any conceivable position in which an oriel construction could look better.

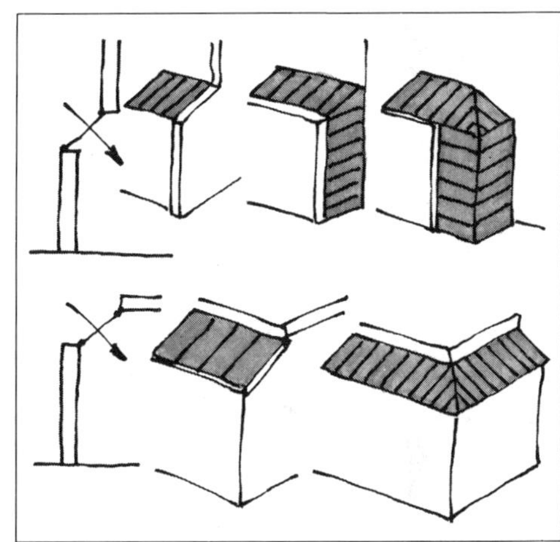

47

Offset Oriels

The values of design and function are equally represented in this chimney corner. From exactly the same sitting area one can enjoy the view during the day and the fire at night.

The arrangement of fields of glass on the edge of roof and facade is particularly unusual and original.

The oriel's effect is not to be enjoyed only from the exterior; it is expressive and persuasive inside as well. During planning, the builders obviously thought equally about internal and external effects.

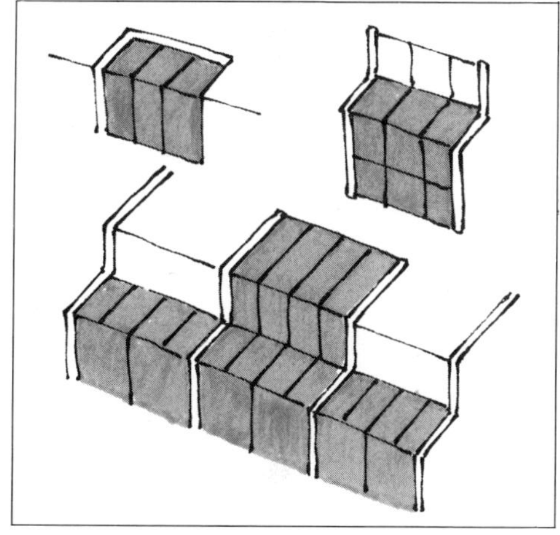

49

Oriels in the Eaves
A Binding Element

As the examples on the preceding pages show, the arrangement of windows in the area of the eaves makes for unusual effects. However, the expressive use made of oriels in this building is particularly clever. Rainwater and snow will build up behind these oriel exten-sions. It must have been at the insistence of the planners that these difficulties were accepted.

Expensive technical solutions—such as special insulation or heating elements that melt snow and ice—were considered justifiable in the interest of originality of design.

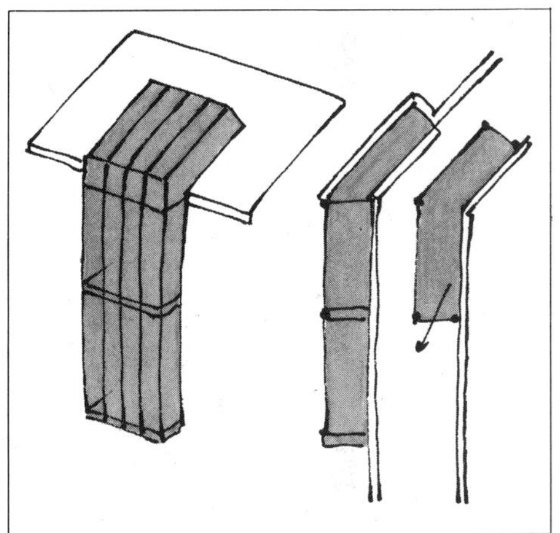

A Sculptural Element

Oriels in the eaves, juxtaposed with skylights and sliding, recessed roof windows, contribute to ventilation. The connection with traditional roof dormers, which open to receive goods and furniture (as at above right), rests on concessions made in modern buildings to enable them to blend in with nearby historical buildings.

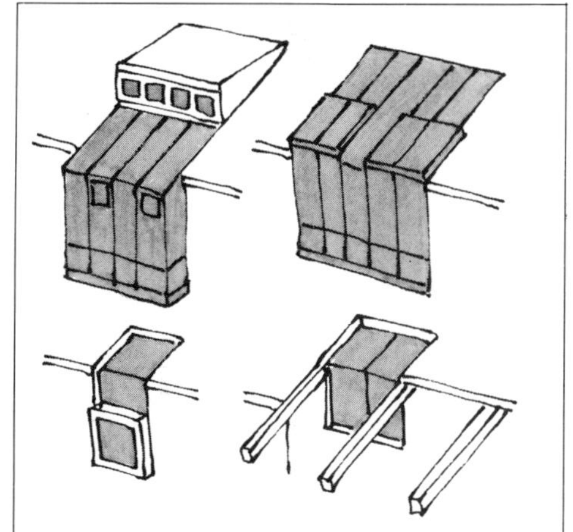

53

With Dormers

The facades of corrugated siding are black; the aluminum of the latticed windows is anodized green.

The cleverness of this particular contribution lies not only in its very sculptural form, but also in the particular combination of materials and colors.

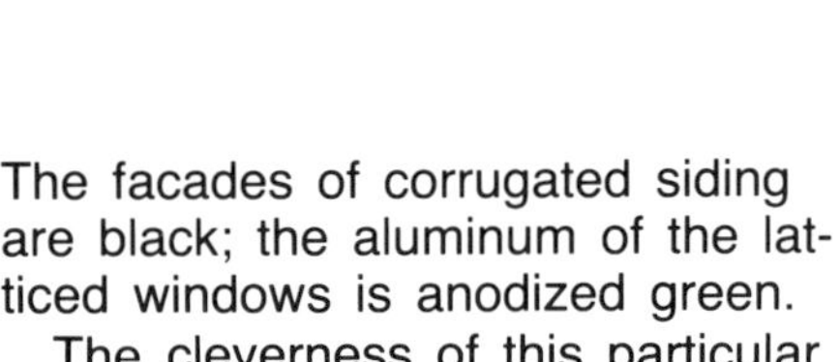

This glassed-in entry passage not only protects apartment entrances from excessive dust and dirt, but also diminishes street noise. This is an especially striking solution, in terms of both design and construction, and could well be applied to the modernization of older buildings.

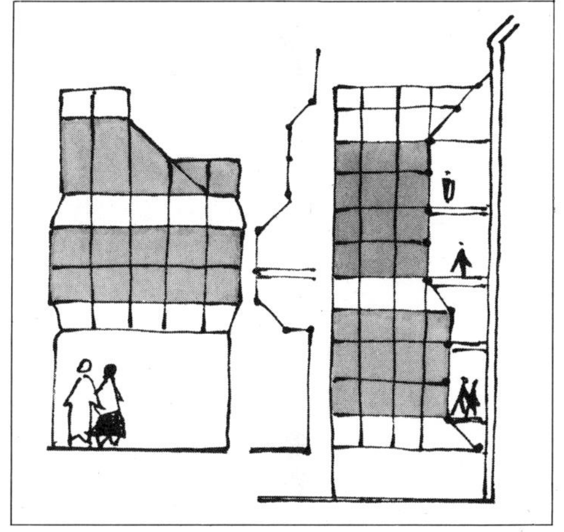

Wraparound Oriels

Through the use of oriel constructions—single and multistoried, resting on ground level or supported by colonnades—these three Renaissance city halls in Bremen, Paderborn, and Lemgo were transformed and extended. It is interesting for us today to observe how structures of brick were overlaid with strongly contoured and painted sandstone extensions, which represented the most modern taste of the period.

The old buildings were ruthlessly renovated if judged to be out of date; the customary practice was to tear them down and negate their very existence. Without reflection, changes were undertaken to implement the preferences of a more modern age. The way these changes were achieved marks these three examples as among the most important constructions of the Renaissance in northern Germany.

The interest in historical preservation in our time forbids the use of such drastic renovation measures; old things are deeply appreciated. However, the question arises: can new forms come into being without

similar practices? Still, old buildings are now too rare for them to be razed in the interest of new constructions. Our achievements lie in the direction of combining different architectural forms, with respect for the old and with appreciation of the new. (Compare the additions to old buildings on pages 116 and 136.)

Sitting places are provided in the hallways of this school by using oriels that overlook the glassed-in sheds that cover the rainy day playground. Red bricks frame the green painted window frames of the oriels.

Outside, the combination of materials is supplemented by the addition of poured concrete surfaces, contrasted with slate-covered walls.

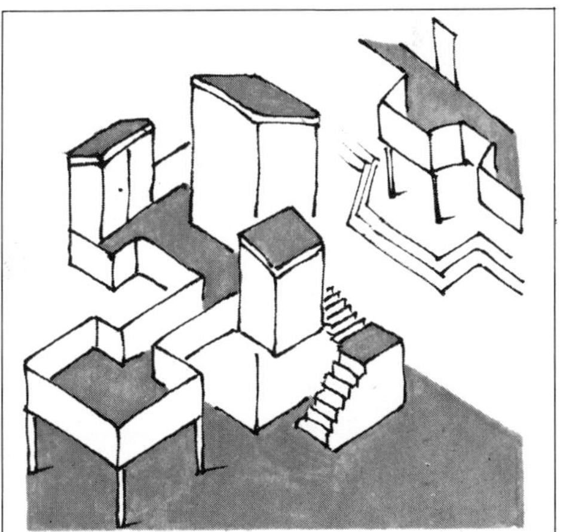

This sleeping loft on supports and the bathroom over the kitchen area are reached by a gallery featuring built-in closets which are hung out like oriels.

The hallways on this upper story of a school extend outwards over a communal meeting hall below.

Interior Oriels

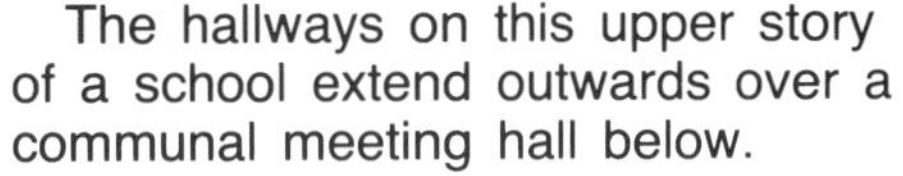

The entrance to the glass-covered reception, residential, and stairwell area is spanned with an arch-like alcove of acrylic glass. The glass domes are supported by light, open pillars.

61

An unusually exciting visual effect is created by combining oriels, crowned by encircling galleries, with raised, tribunal-like sitting areas, overhung by planted balconies that integrate smaller, bordering rooms into a spacious, covered hall.

Courtyard Oriels

The free-hanging stairwell builds an internal oriel overlooking the giant acrylic glass oriel that extends the multipurpose room. (Compare with preceding pages.)

This tiny school courtyard not only provides light to the different floors, but also contributes to an overall cheerful quality. Oriels are built in on the ground floor. In the upper floors, the glass roofing creates a similar effect. The intense red of the window latticing contrasts with the green of the plants. This is an unusual combination of oriels, creating both conservatory and courtyard.

With Conservatories

On very tall buildings, the upper story is frequently used to create interesting architectural effects. Thus, oriels are often aligned to form enclosed rows or extensions, or are used as isolated elements.

The monumental, mountainous forms of large apartment complexes sometimes seem to consist only of oriel forms thrown out in all directions.

Buildings made from prefabricated elements do not have to compare badly, in design quality or originality, with conventionally designed and constructed buildings. As these examples show, they can even surpass them.

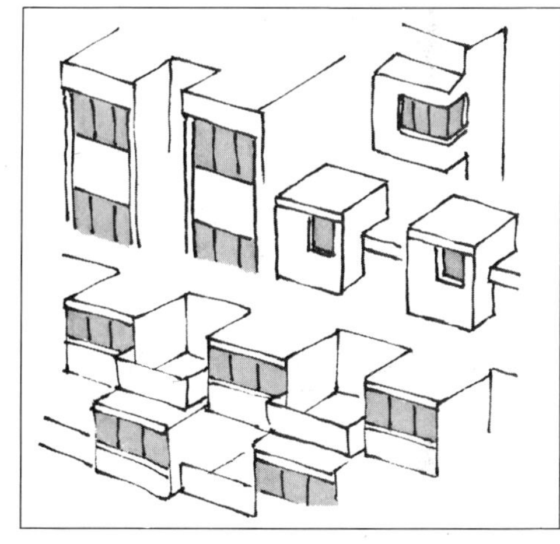

65

Oriels in the Upper Story

The stacking, adding, and grouping
of different building elements
creates structures that seem to be
made exclusively of oriels.

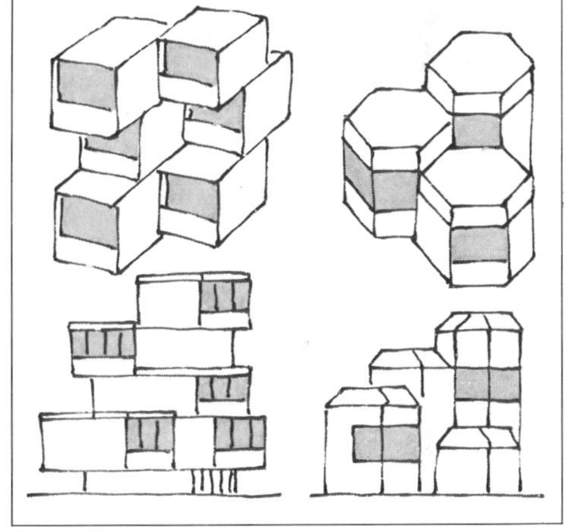

Oriel Constructions
Cubes and
Honeycombs

Building elements that are set at right angles or slanted, or stacked high in the air at angles to the main structure, produce of themselves oriel-like structures.

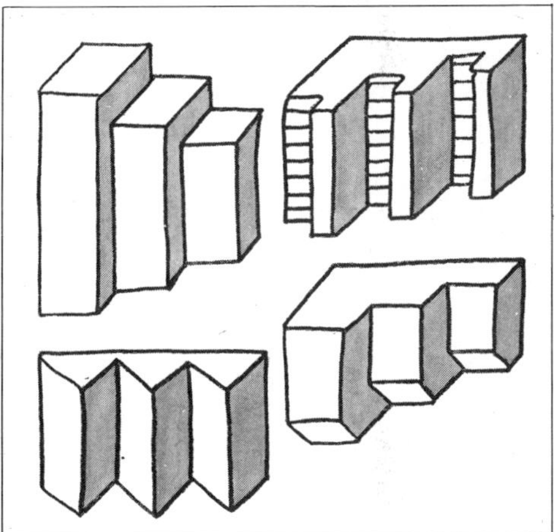

69

Jagged

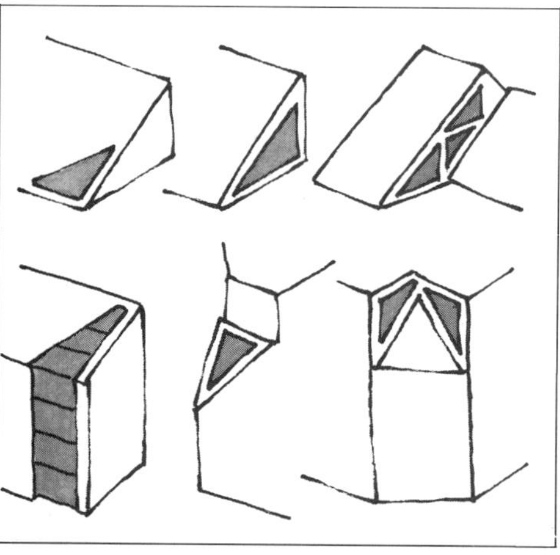

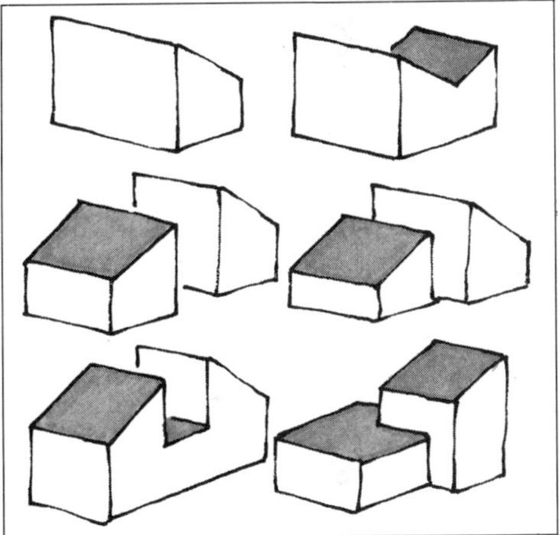

The use of lean-to roof lines, aligned with or against one another, staggered or extended out at different distances, creates a variety of architectural effects while remaining free of complexities. Whether by careful planning or by chance, numerous triangular forms are created by combining different-sized roof elements with different architectural elements. Whatever the original intent, these forms are very effective.

Triangular and Pyramidal

Constructions like this, despite their sculptural aspect, often have a very classical floor plan.

Two rectangular wings, laid diagonally over one another, create a conjunction of points that forms a guiding theme for the floor plan (which must not, however, be taken as too rigid a directive). The corner angles are prescribed, but not the height of the eaves.

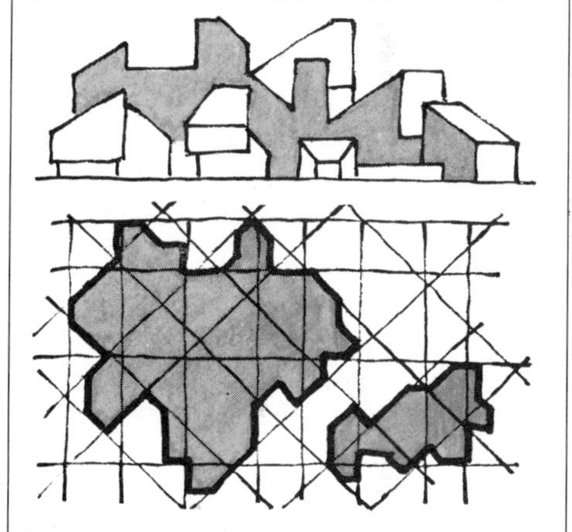

Free-Form

Construction elements such as oriels, balconies, loggias, and lean-to and stepped roofs can be combined so that the total effect is more nearly that of a sculpture than that of a house.

An exciting experience: with each change of view, the exterior and interior spaces metamorphose to produce new impressions.

Today, there is a renewed respect for buildings in the style of Art Nouveau or those built at the beginning of the earliest industrial expansion. Oriels, dormers, balconies, and towers of these periods are being restored rather than torn down.

After a period of considerable architectural austerity, the wealth of forms and colors possessed by these buildings makes a favorable impression. They no longer affront our attitude of conceptual coolness, but rather challenge us to emulate their creativity and diversity.

75

Corner windows which either extend out or cut into the interior space can, if they are aligned singly or in bands, be considered oriels and treated under that rubric. Corner windows differ greatly in height: they can be as high as the room itself, or they can be mere sky-lights. They appear more oriel-like when the window surface is broken up by lattice work; with large glass surfaces and support walls, they seem more like simple windows.

Corner Windows
Rectangular

Both of these different corner windows have in common their glass-to-glass-edge solution. Above, heavy concrete and, below, thick wood fulfill the same structural function. Building corners which are constructed so as to carry weight in an unusual position create an effect that is pleasantly bewildering.

By their position on the corner edge of a building, corner windows can be classed in types, according to whether they are convex or concave, and whether they are positive or negative as seen from inside or outside. Because they are hemmed in deeply on both sides by structural elements, corner windows must necessarily be counted among oriels.

Windows and filled-in walls on building fronts or on corners can be sculpturally emphasized. This, in turn, structures the design of the facade without reference to structural necessity.

Sculptural

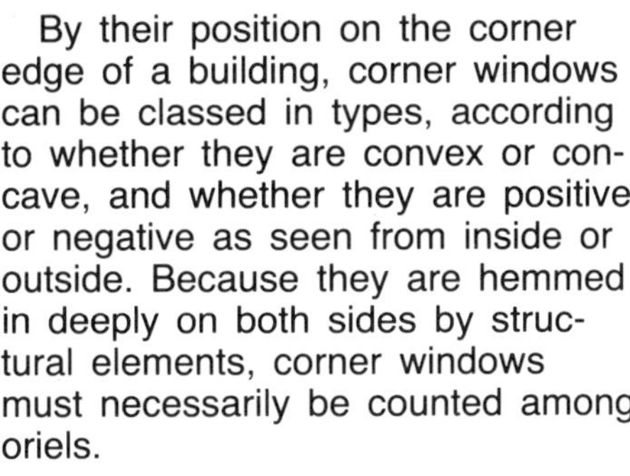

Curved window surfaces—either vertically placed on building corners or horizontally placed, as in cupola-like glassed-in areas seated in the roof eaves—allow a view similar to that of oriels, they also resemble oriels in structure.

It is rare to find angular corner windows that are perpendicular and stretch from the filled-in wall to the floor above. Generally speaking, they result rather from two building edges being bluntly joined together. No matter how they are made, the result is interesting building and window forms.

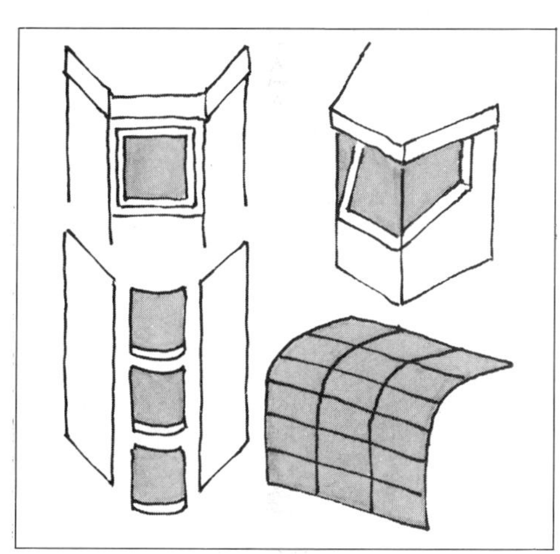

Rounded and Angular

Glass facades that are staggered—outward, inward, or in combinations—often fulfill functional demands admirably, as in this example: a machine warehouse. In other cases, the design aspect takes precedence, as in this counting-house of a bank (right).

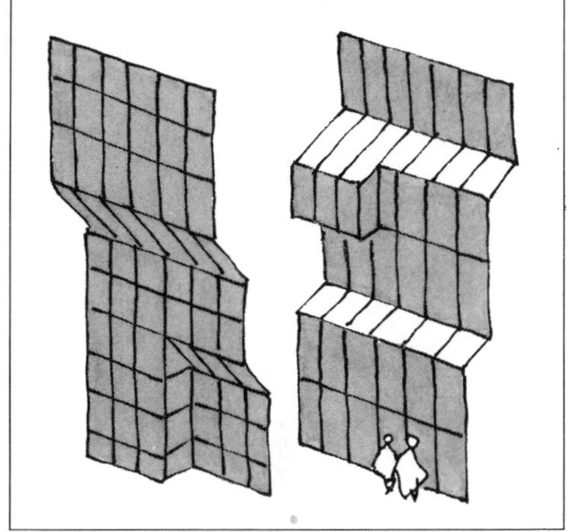

Sculptured Glass Facades

In the construction of apartment buildings it is rare to see windows set at a slant. These examples are exceptional and therefore particularly interesting. On the ground-floor level, there is a community pool that serves the apartment block. In such situations, the use of slanted windows is increasingly prevalent.

So-called roof windows, which either open outward or slide sideways at roof level, are the most well-known variety of slanted window. They are often used in pairs, next to one another or one over the other.

85

Protection from intense sunlight is a major consideration in the placement of slanted windows. The example here shows a slanted window shade.

Slanted Windows

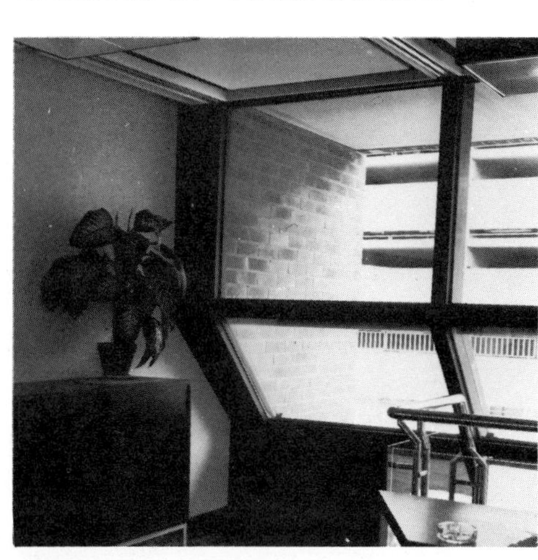

These are work alcoves and seating arrangements illuminated by a glassed-in lean-to wall. A large light shaft lights a dividing corridor (above). A specially-designed ladder wagon allows window cleaning in perfect safety.

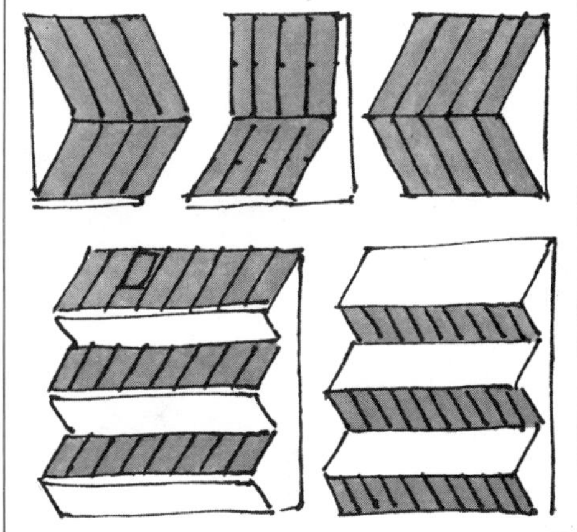

Other reasons for using slanted windows are to hinder distracting reflections and to decrease surfaces that would be affected by wind resistance. Architecturally, folded and slanted surfaces are attractive.

These are some methods of arranging glass surfaces:

Like a roof surface tilted upward; this allows better illumination for studio windows, but collects considerable dirt and dust.

Slanted inward and downward; this offers a better view, and very little dirt collects.

With both slanted surfaces coupled.

With one surface slanted from an already perpendicular surface.

Recessed Windows

Old-style balconies, loggias, and conservatories, either with the original glazing or subsequently modified, can be counted as oriel constructions. These examples, with their delicately filigreed grillwork, both molded and hand-smithed, evoke a special response in us today, when a highly differentiated profile does not go unappreciated.

Balconied Oriels

Here are variations in the combination of balconies and oriels. One- and two-story elements are either railed-in, glassed-in, or left open, contrasting delightfully with the large, flat, closed wall surfaces. The floor plan corresponds to the elevation; both show honeycomb projections.

Balconies with strictly or freely modified multiangular designs, arranged singly or connected by tie-beams, produce lively facades.

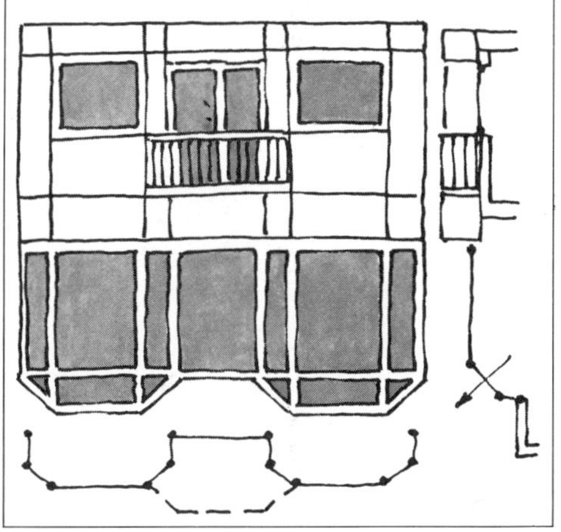

91

Honeycomb

In this apartment complex, the main intention was to provide a differentiated surface. Considerations of privacy, such as being able to sit on one's balcony out of the view of neighbors, were neglected.

The use of vertical supports relates to various aspects of design: function, style, and construction. Plants can be trellised; individual balconies can be closed in with latticework or with different kinds of partitions, which are simple variants.

Balconies provide open-air seating, either enclosed on three sides, as loggias are, or cantilevered out into space. They are included in our inventory because of their similarity to oriels. They are arranged on perpendicular facades singly, in groups, in rows, or in combination with step-like elements forming a kind of mountainous building mass.

They are usually partitioned at waist height to prevent drafts.

The sides are often even higher to allow for privacy. However, there are commonly structural reasons for this, too, especially in the case of prefabricated constructions.

These constructions differ in the way their weight is carried: either by walls or by foundations.

Many different factors are involved in the planning of balcony construction. The location of the balcony is dependent on its exposure and on the kind of interior space that abuts it. Placement can be on gable ends, on the facade, in the eaves, or on the corners of the building. In the floor plan, balconies can be rectangular, quadrilateral, polygonal, or round. The roofing is usually provided by the balcony above.

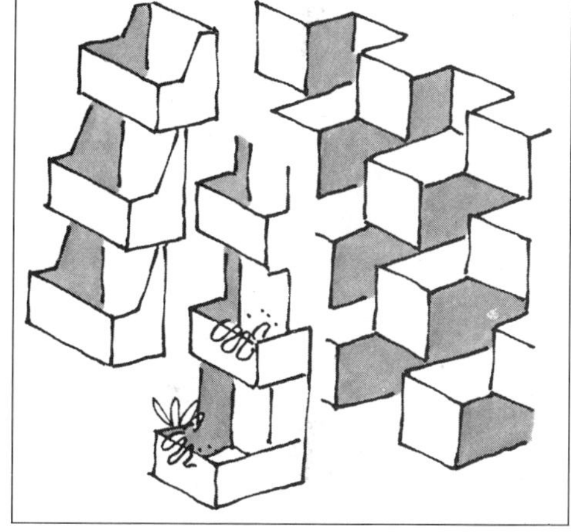

93

Balconies rest on supports or cantilevered "arms." They can be built on top of the roof of a more extensive lower story, or they can be slung between walls.

Different furnishings can be provided in the form of flower boxes or small closets for garden furniture. Tenants are responsible for providing their own protection against wind and sun.

Stacked

Balconies in conjunction with oriels permit a view both outward into free air and inward toward the living space.

Plantings and the amount of protection from sunlight vary according to individual taste. The result is a cheerful and diversified, human-scale mass dwelling.

The building here is so covered by oriels, coupled with balconies, that the load-bearing elements are completely hidden, except at ground level.

Triangular balcony forms are relatively new and are often preferred because of the distinctness of their effect. We can compare different solutions: closed and open, light and heavy, symmetrical and asymmetrical.

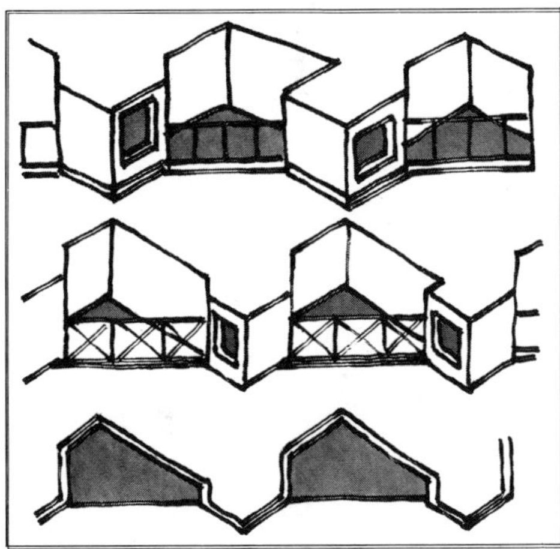

95

Triangular

Window boxes are increasingly employed, not only in gardens and terraces, but as a construction element on balconies and on sculpted window units on the facade. In order to provide the essentials for plants, one must take all their biological needs (in addition to the purely architectural aspects of construction) into account in the design of window boxes. This calls for a great deal of creativity and careful planning.

In these examples, construction and costs were balanced.

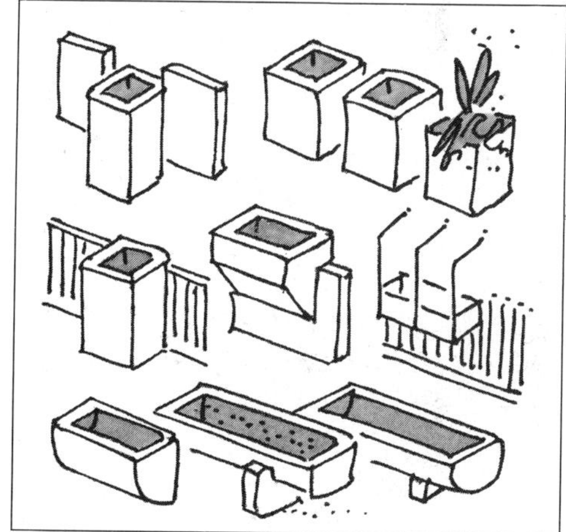

Window Boxes

Buildings can be joined by using a variety of constructions, such as: glassed-in walkways that lead over roofs; suspension bridges that span the interim space; or buttressed bridges or breezeways. Angular, arched, and semicircular roof forms have all proved their worth. The glazing, with glass or acrylics, is usually permanent.

Breezeways and Bridges

101

Dormers

Dormers go beyond the merely functional. They not only provide light and ventilation to attic rooms, but also serve to decorate them. In every period of traditional architecture, they have been represented in a wide variety of styles.

The type of dormer that is chosen often depends on how and where it is placed on the roof. Is it in the eaves or high on the ridge? Does it grow out of another construction, or does it result from the stacking of several architectural elements?

Recessed lean-to roofs make it possible to have skylights facing in several directions.

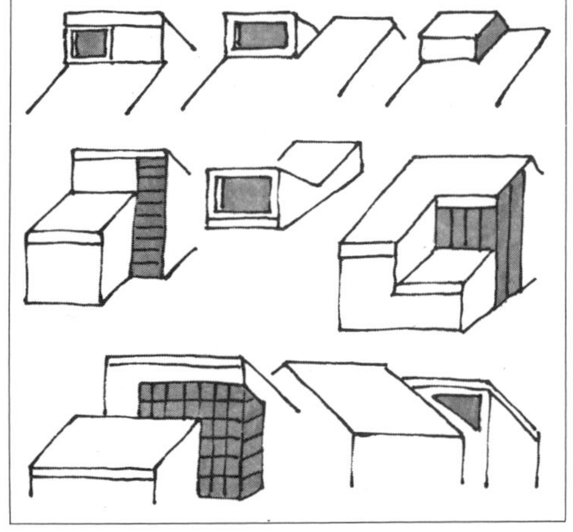

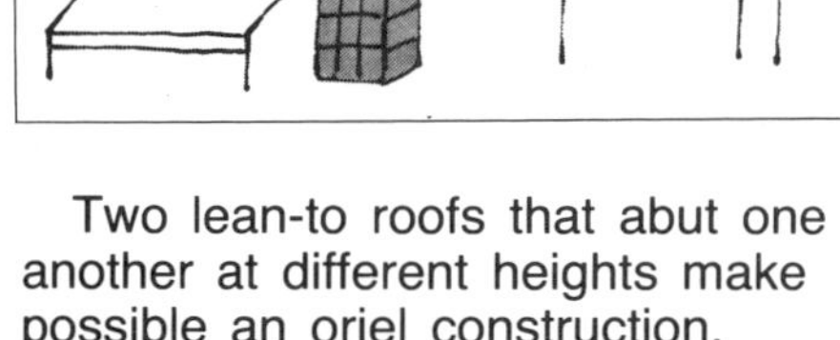

Two lean-to roofs that abut one another at different heights make possible an oriel construction.

Building elements that extend beyond one another on the sides offer other opportunities for oriels. Triangular window spaces often occur when different rectangular building elements are combined.

Modern dormers are distinguished from dormers of earlier periods by their mode of construction and their style.

Examples include trailing dormers with roofs paralleling the actual roof; flat, box-like dormers with more traditional pitched roof gables; and hipped-roofed dormers with round or oval openings (so-called "bullseye windows").

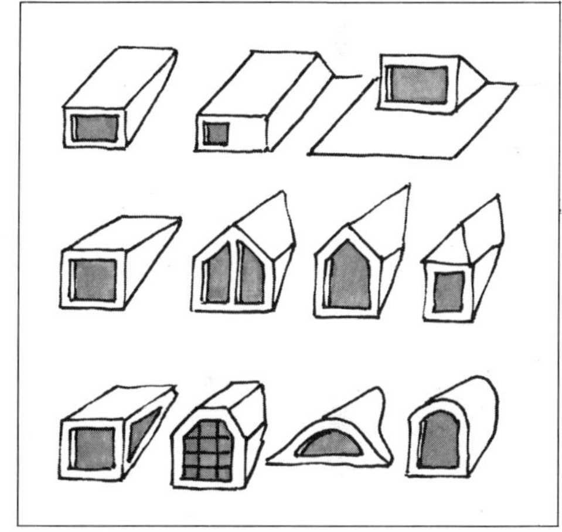

When we consider the variety of placements possible on building corners, the permutations seem endless. Equally varied are the possible arrangements of window openings and the different possible types of fenestration.

103

On the Roof

Dormers—unlike some oriels—must have windows that can be cleaned from inside. A few rare examples permit outside washing, as long as they are within a short distance of ground level. In the case of oriels, window washing is done almost exclusively from the outside.

At present a wide variety of options exists for glazing: with or without mullions, and with large or small windowpanes. Also popular is the practice of opening the dormers on the sides, affording a view over the roof tops.

Above and beyond the functional are the design aspects of various illumination options.

In the case of sculptural oriels extending out from the facade, the variety of lighting designs far outstrips those available for standard windows.

Here is an example of a conference room lit by sunlight entering through slanted windows that are the roofs of triangular oriels.

Current architectural practice makes frequent use of pitched roofs and gables; roofs have become bodies of architectural interest rather than simply lids for dwellings. Sometimes roofing material is extended to cover the sides of the facade. Rain gutters are hidden behind the exterior covering or skin of the building.

On Gable Ends

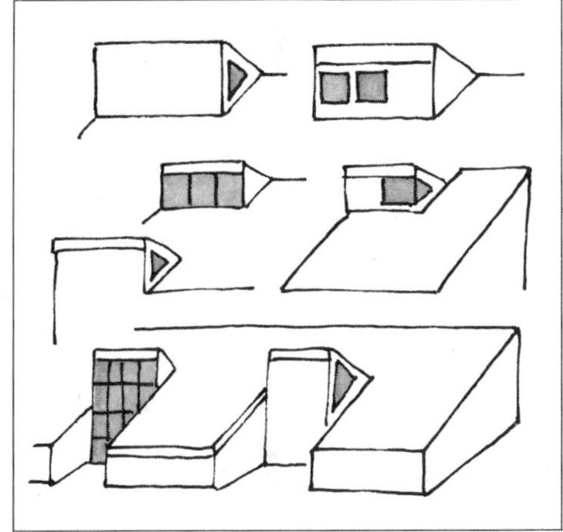

Roof extensions whose own roofs slant in the direction opposite that of the main roof offer enormous spatial gains for attic rooms and allow extensive views.

Although window openings are most commonly only frontal, it is not unusual to find them on the sides. Joining these dormer forms to the main structure, although difficult, makes possible a rich variety of realizations and is well worth the effort. Naturally, rainwater and snow collect in the declivity between the main and dormer roofs.

These points of connection must be built with the utmost technical precision. Insulation foils or sheeted lead prevents leakage. Heated coils of wires can work to thaw ice build-up.

Facing the Opposite Direction

It is possible to differentiate between rectangular and quadrilateral floor plans and among straight, slanted, and swelling elevations. Illumination is effected from above, from the side, and from the gable.

Glazing can be of single- or double-paned glass, tinted or wired, with or without mullions.

The arrangement of skylights can be on pitched, lean-to, or flat roofs; it is most impressive on the more unusual "tent" or swelling roofs. The inclusion of skylights in designs for individual dwellings is as frequent as in those for public buildings such as schools, kindergartens, and nursery schools.

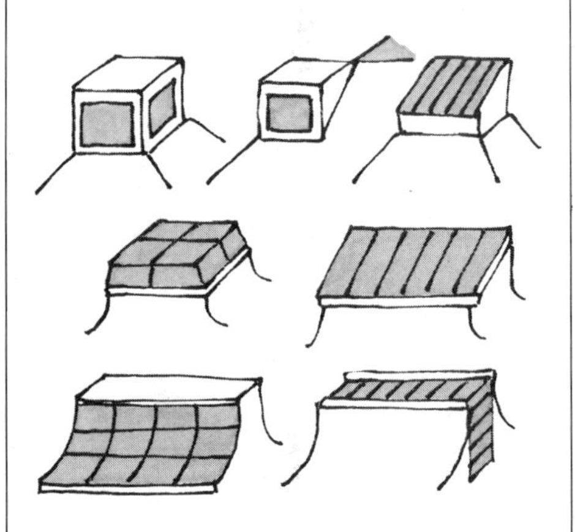

109

Skylight Constructions

The form of the roof was often saw-toothed or arched; exceptionally, as in this case, it might be rectangular. Today, lightwells may be used as independent constructions or sheds on a normal roof construction, or even flat on the ground. The lightwell construction has seen wide application.

Lightwells were developed for industrial structures. Nowadays they are transformed and applied to a variety of architectural tasks.

Originally, lightwells were made by filling in half-timbered hall constructions with glass instead of masonry.

A distinction can be made between lightwells based on pitched, lean-to, and flat roof constructions and ones that take rounded or arched forms. Their placement can be perpendicular or tilted, swelling, unfolding, or projecting from the main roof body.

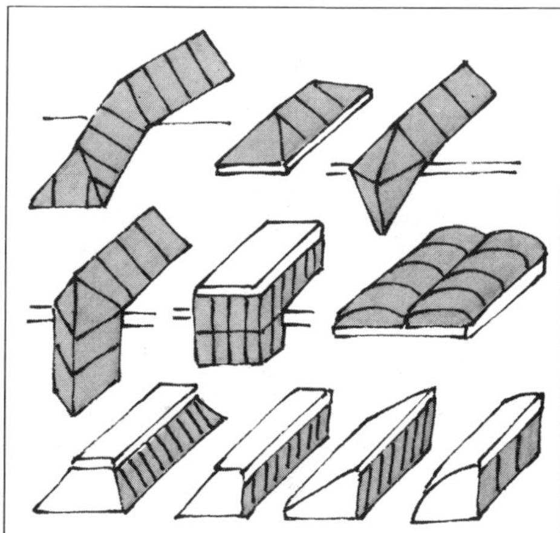

Glazing can be permanent or can be provided with transoms for ventilation. The light can fall from above, from the sides, or frontally. Often a combination of these is found. A northern exposure ensures consistent illumination unaffected by glare or shadows.

Glass Sheds

Skylight domes are the only kind of skylight that can be manufactured in one piece.

They were immediately accepted by the construction industry as a useful prefabricated element. Cupolas can be flat or arched, rectangular or circular, with or without mullions.

Pyramids with triangular or quadrilateral bases exist as well.

The extremities of the cupolas may be perpendicular, slanted, or bowed. Acrylics serve as glazing material. There are also conventionally- or custom-designed skylight cupolas available. These often have mullions. The prefabricated synthetic domes or cupolas are completely closed and cannot serve as ventilation sources. Domes of double glass exist and provide excellent insulation.

113

Skylight Cupolas

Glass roofs often cover entire court-yards. They may illuminate halls, auditoriums, conference rooms, city hall entrances, or cashier stalls. Their forms are based on technical demands and fulfill particular requirements.

Glass roof constructions must be designed not only to bear their own weight but also the additional weight of snow. Usually, they span the hall without the aid of supports, but this must not interfere with their illuminating function. High, open metalwork constructions result from this design, and their appearance must be carefully considered. Inside, the load-bearing construction

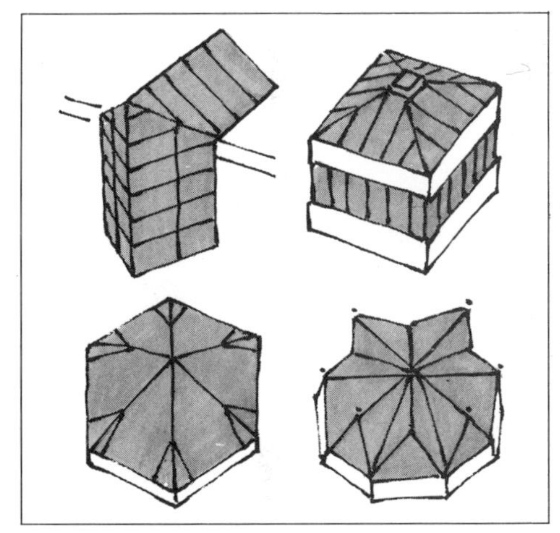

is often altered by the use of modular steel structural components that are themselves glazed. This prevents dissipation of heated air and decreases heating costs. Lighting and air-conditioning elements can be easily mounted above this transparent substructure.

Architects strive to apply this attractive exterior feature of glass roofs in various ways: for example, in church architecture, as spires; or, as in the case shown here, in schools, to add an eye-catching feature.

Construction requirements, costs, and available materials dictate the solutions for glass roof constructions that planners should choose. Consequently, the most economical construction elements are those most often used.

Glass Roofs
Common Forms

Added-on superstructures are a frequent practical solution to the problem of extending old buildings.

New and old architectural features are combined in varying degrees, sometimes even when a project is begun with the intention of building a consistently modern structure.

The example below displays consistency with older constructions. The above solution is animated by contrasts of form, colors, and construction materials, leading to the separation of functional and technical installations. Left: The air-conditioning units are mounted in the open, under the glass roof.

Pitched

Large building projects demand unusual solutions. Both the quality of the project and its mass must be considered.

In the past, public buildings were expected to manifest a certain level of impressiveness; large portals were considered absolutely necessary. Today, architects try to offer the client an element of human scale.

The recess hall on the roof of a school (left) can be understood in this light, as can the high school stairway at right.

In the first of these constructions, the architect has opted for an open, tube-like construction, roofed in glass; in the second, a wooden roof was selected. In the first case, the design of the roof was integrated into the construction, while in the second the construction was simply capped with a roof.

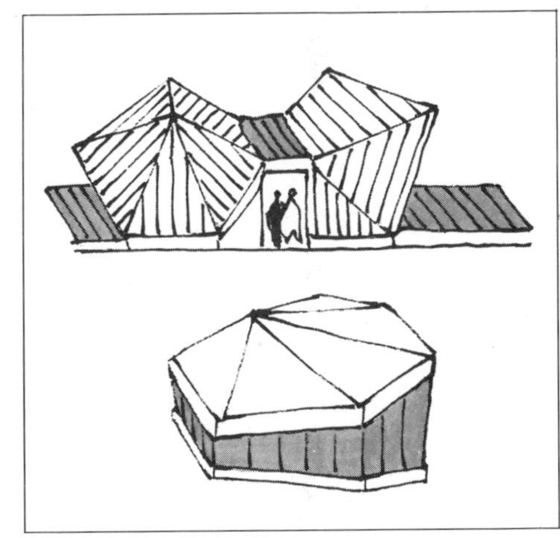

119

Free-Form

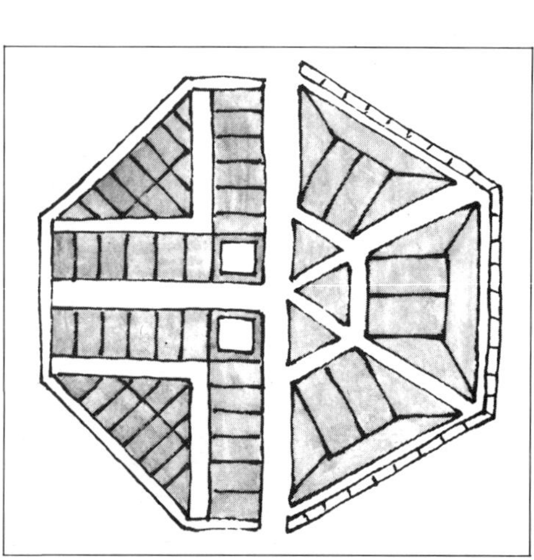

Glass roofs furnish light to interior courtyards or serve as skylights for interior spaces. The sculptural variations can help control drainage or strengthen the construction itself, and they have an equal effect on the internal and external appearance of the structure.

Vertical plate glass tends to remain cleaner than slanted plate glass. So-called "lantern" constructions are preferable as a result. Glass roofs give a building an ornamental character when viewed from taller neighboring structures. A "fifth dimension" is achieved with a view from the roof.

In this example, a planted light-well forms part of a cafeteria that is unusual in conception and lighting. The "lantern" top has ventilation transoms. Thermally-activated blinds control the amount of sunlight that enters.

121

Over Interior Courtyards

Porches serve entrances as shelters against the elements. In addition to this function, they can accent the entrance. Sculptural extensions create interest and provide good construction supports for other facade extensions. One can differentiate between angular and soft forms, closed and transparent forms, and monochromatic and polychromatic variations in the construction materials: wood, steel, aluminum, synthetics, and glass.

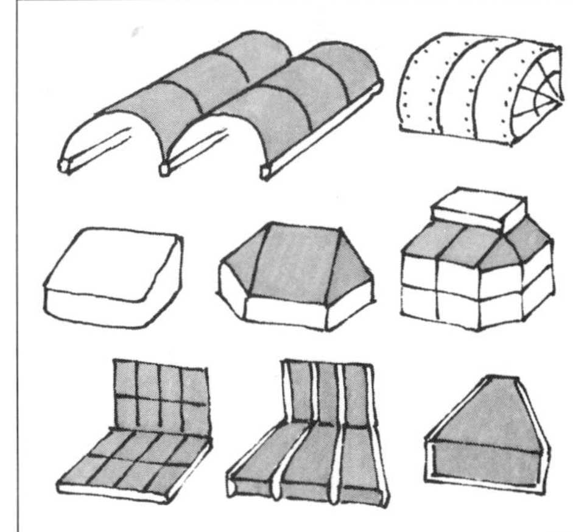

123

Porches

Two large oriels, with seating, extend directly into the exhibition space.

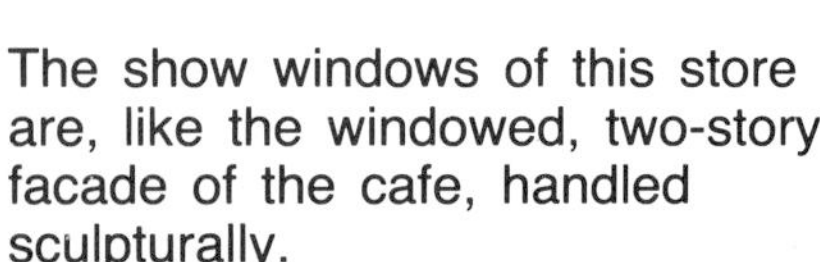

The show windows of this store are, like the windowed, two-story facade of the cafe, handled sculpturally.

This museum is rendered unmistakable by the use of the fine leaded-window fenestration, painted red. The museum and the cafe are unusually well bound together optically in this architectural composition.

Strength and distinctive form are evident in these show windows that stretch from the floor to an indentation situated two glass divisions from the ceiling and then on up. The sidewalk seems to be drawn into the building. Radiators stand freely in the oriels, and air enters through slits high up on the slanted glass surfaces.

Show Windows
Exceptional Forms

These examples of unusual show windows testify to the richness of variety possible in this architectural form—and it should be borne in mind that most of them are not even designed by professional architects. These free-form windows are difficult to realize on paper beforehand and require careful implementation.

Extraordinarily slender oriels are made possible through the use of steel framework. The tensile strength of steel is very great, and connections can be of very small dimensions. Joints can be screwed, molded, soldered, or even glued.

The possibilities offered by painting are numerous; rust can be controlled by galvanization. One disadvantage of steel is its temperature conductivity, which often causes significant heat loss. However, this point should not be overemphasized, since the steel frameworks themselves are often very slender.

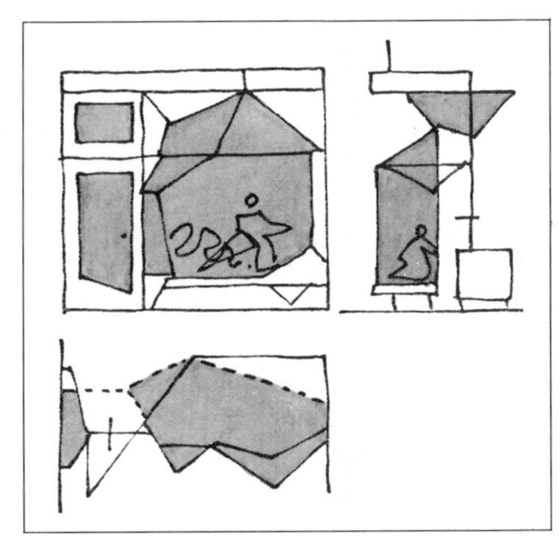

127

Free-Form

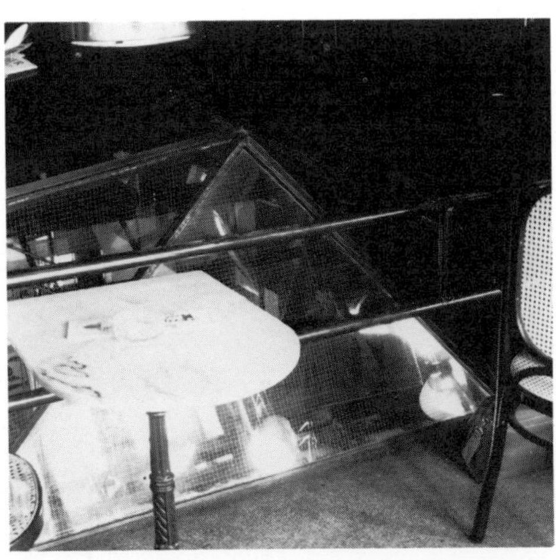

These striking examples of modern plate glass oriels, constructed within the old brick columns and arches, prove that it is possible to combine modern architectural features with existing buildings without being overwhelmed by insurmountable difficulties. Good architects welcome this opportunity to work with given parameters to create a persuasive blend of old and new. The old high-ceilinged shops were divided to make a cafe entresol and a passage that fronts boutiques and shops.

Oddly enough, the impetus for this inventive remaking was the widening of the street, which necessitated putting the sidewalk under the columns.

The old iron column, with its molded capital, supports the double T-beams and the masonry filling. The new show windows and oriels, framed in steel, contrast with the poured concrete ceiling and brick walls. There is similarity in the construction methods, but a strong contrast in the building materials used.

Additions to Older Buildings

Showcase windows and vitrines are used and constructed in a variety of ways. They often are built to hang on the facade, or on columns or pillars, and are bound to the building as a whole. Since they often protrude or extend from the facade, they can be classed with other extensions and ultimately with oriels.

Although they cannot strictly be counted as oriels, vitrines are worth considering here. Not only do they share some of the same constructions, but, with their free-form sculptural elements, they are architecturally desirable for some of the same reasons as oriels.

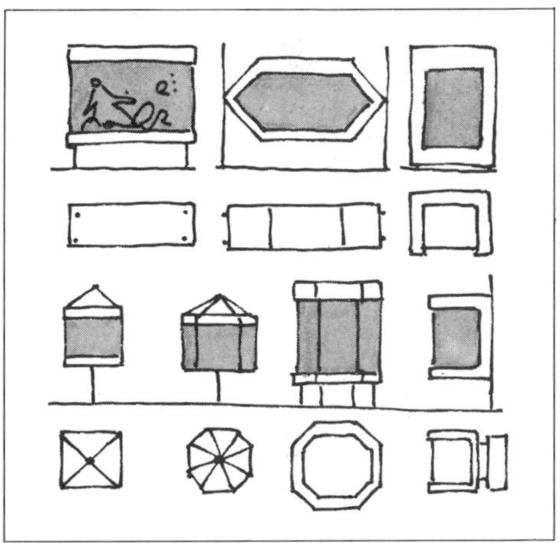

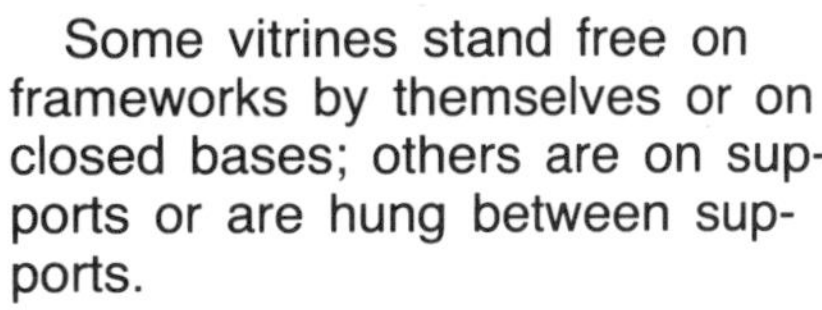

Some vitrines stand free on frameworks by themselves or on closed bases; others are on supports or are hung between supports.

131

Vitrines

Streets are evaluated differently now than they were a few years ago. They are no longer thought of exclusively as the domain of automobiles or as conduits for completing shopping errands.

Automobile traffic is frequently divided from pedestrian traffic now. Easy communication between shoppers is thereby made possible, and a shopping expedition once again becomes a pleasant experience.

Architects are responsible for the planning not only of facades, but also of small squares and passages. These passages are often defined by sculptural building elements. Show windows blend into oriel-like constructions, and vitrines separate themselves from the facade and become almost like pieces of furniture arranged on the sidewalk. Those that can be entered resemble pavilions and are sometimes several stories high. If unenterable and smaller, they more nearly resemble booths or kiosks.

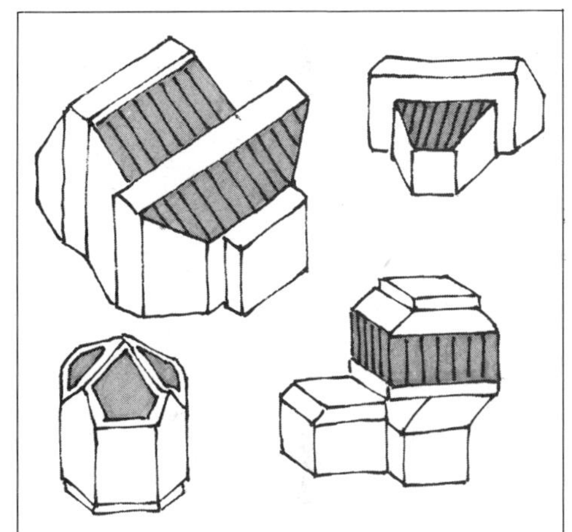

Pavilions

Passages can be extraordinarily diverse in appearance. We distinguish between those that have been added to older constructions and those that are freshly conceived. Passages can be of one or more stories and can be equipped with galleries or bridges, walkways, stairways, and escalators.

Roofed portions of streets protect central shopping areas from the elements and, if they are also closed on the sides, from the cold. The sheltering roof constructions are pitched, arched, or (frequently) barrel-vaulted. This last construction is statically favorable and structurally simple.

Passages

Within the passages, facades and showcases can be designed that are not subject to the normal demands of weather conditions. The walkway is transformed with seating areas, planters, fountains, and even art objects possessing an urban flavor.

The theme of passageways is enjoying renewal in today's extravagant contributions.

As in the past, the combination of old buildings with more recent ones remains an important theme in architectural history. The requirements demand extraordinary solutions and achievements.

Oriels often offer a key to the solution of assignments involving extensions of or additions to older buildings, or even complete changeovers in their functions. It is interesting to note that, in the case of these functional additions, little attempt is made to match the old structure exactly.

These solutions offer exciting contrasts between the old and new building segments, while the integrity of the total design still remains recognizable. The past has a future, and the present finds a new mode of expression.

New Oriels on Old Buildings

Staircases that stand free of the facade or next to the building bear witness to advanced structural or architectural considerations.

It is possible to design a floor plan independent of convention. For example, stairwells can become closed elements that are like vertical axes or accents, placed only tangentially to the different floors.

Structurally, it is important to provide transitional construction between higher and lower buildings. The use of different construction materials and the different load-bearing capacities of two buildings can cause cracks and even fissures to appear. It is sensible to use a stairwell as an effective divisional device. Other elements of the building can also be used in this manner.

In general, the function of an object, such as a tool or a piece of furniture, should be easily recognizable from its appearance. The same demand is made of buildings and individual rooms: the entrance of a building must be easy to find, and the floors and their extremities easy to recognize.

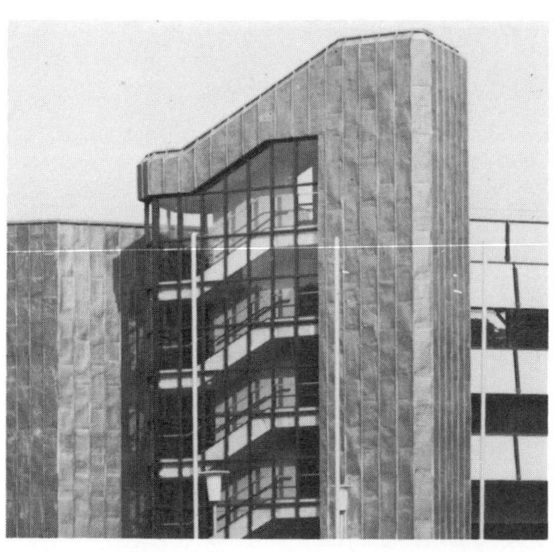

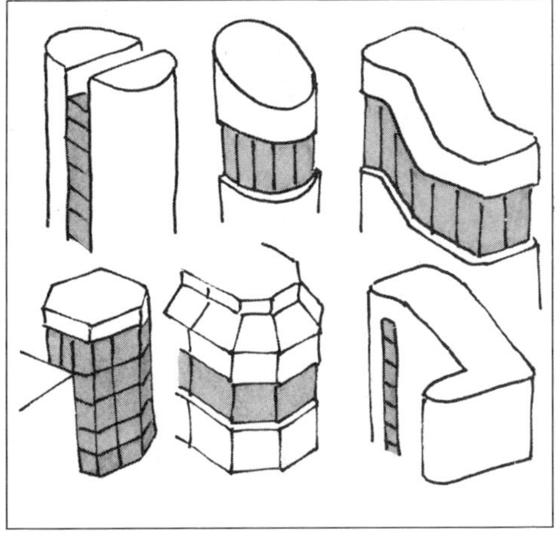

Stairwell Towers
Round

Even the most pragmatic project demands planning considerations of style and design. The architect's task is to encompass important technical and functional components within the framework of artful architecture.

Good architecture is a fine art with the task of fulfilling the stylistic demands of a given period and, at the same time, making further stylistic developments.

These examples of different architectural achievements all aim to attain the highest standards of architecture.

It is an ancient conceit to construct stairwells so that the size and height of the individual steps can be seen from outside. Many examples of this from the history of architecture come to mind.

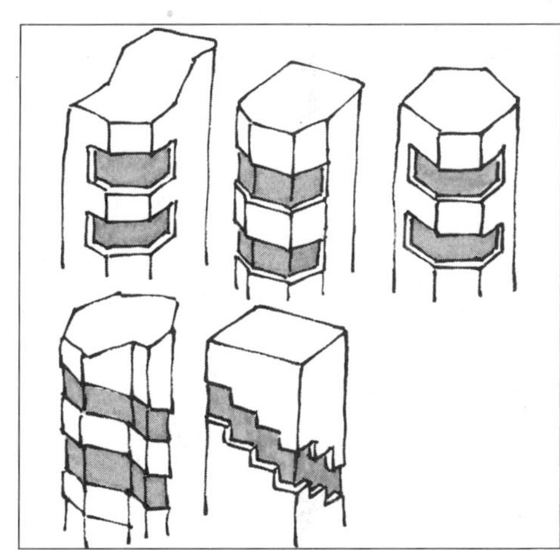

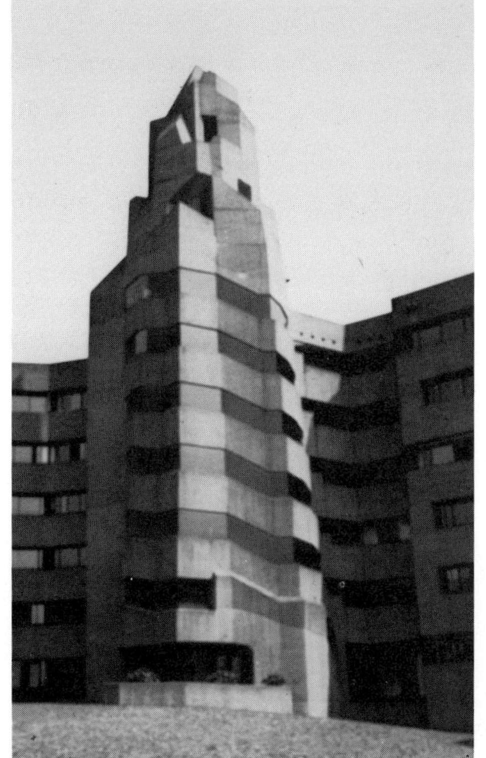 

Polygonal

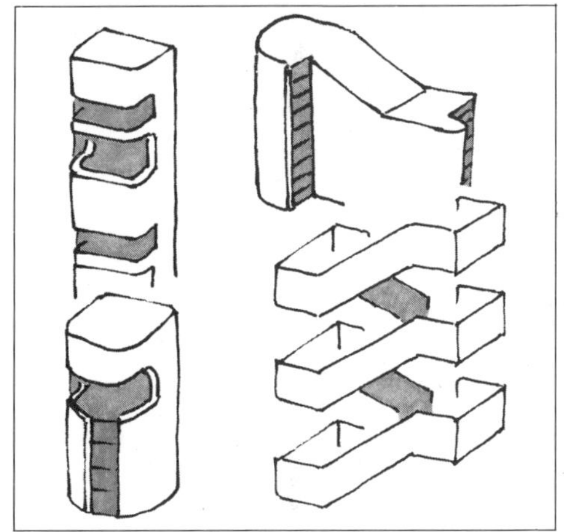

There are practical reasons for placing stairways outside the perimeter of the building.

In the form of open emergency stairways or fire escapes, they fulfill code requirements and provide escape routes; but stairwells that are enclosed also fulfill these requirements, without disturbing the overall design of the facade or making the users anxious.

Considerations of both security and architectural design require good architects to take pains with their proposals. Few would voluntarily opt for a fire escape in the form of an exterior staircase, which, as a large sculptural element, can seriously disturb the design of the facade.

Curvilinear and Stepped

Hanging the stairwell on the outside walls is a practical and space-saving device. The idea is original in conception and construction. Glazing in stairwells provides protection from wind and weather, and makes use of the stairways safer and pleasanter. The facade receives an extravagant note with the addition of oriel-like stairwell constructions.

Glassed-in, conservatory-like arcades offer an alternative to the usual solution of closed-in stairwells that connect the floors like an extensive pulpit.

There are open and closed arcades, and arcades that are like balconies or bridges; some are located on facades, others on roofs. Stairwells can be between or in front of buildings.

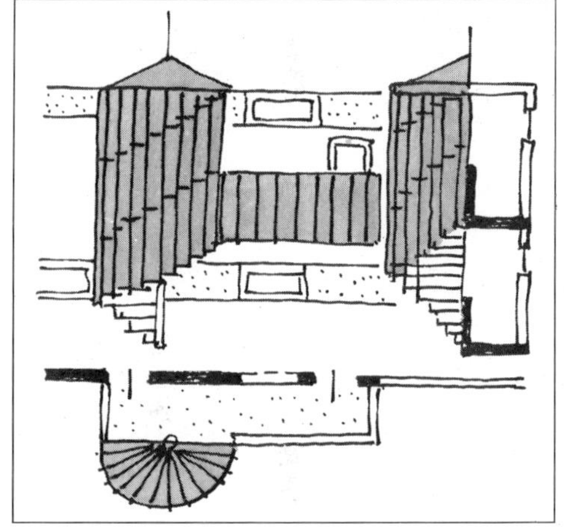

With Arcades

From this arcade, a small winding staircase leads to a second, smaller arcade passage, leading in turn to additional apartment units.

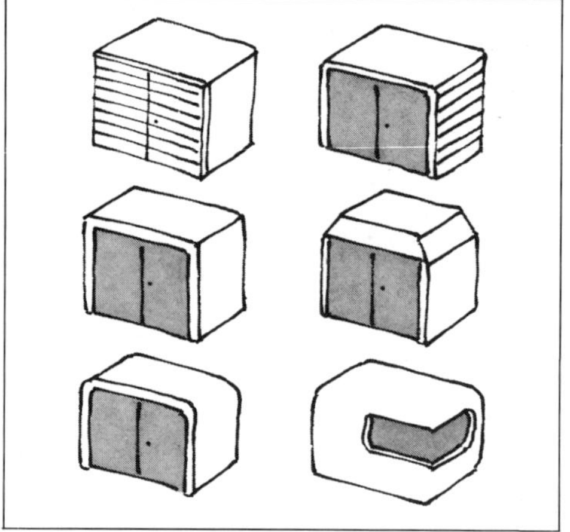

There is a great variety of possible designs for oriels that are used as wind cheaters. Along with angular, box-like, or carved forms, there are also those with rounded or sculptured contours.

These examples are also distinguished by the choice of materials for construction and facing. Heavily colored materials contrast deeply with those left in their natural state.

Entryways that stand as independent structural elements in front of the facade are optimal solutions.

They serve the same function as porches and accent the entrances. They are outside the ground plan and can be freely designed independently of it.

In the examples shown here, each entryway either is carefully blended with the building's style or stands in contrast to it.

Hexagonal entryways, particularly those that are fully glassed-in or that have angular roof elements, can easily be included in our discussion of oriels. Painted, enameled, or anodized steel- and aluminum-framed constructions are especially elegant.

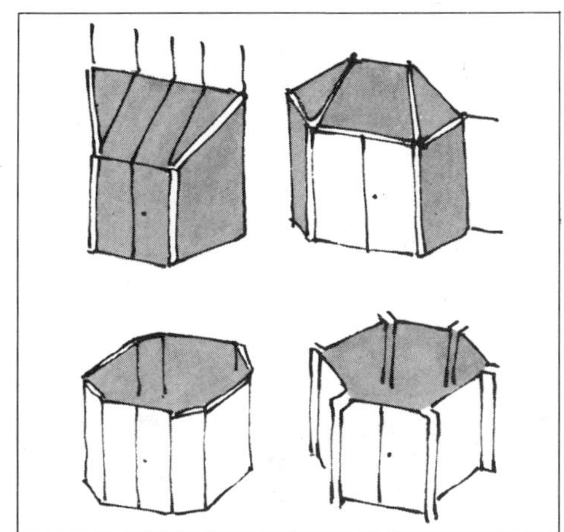

147

Entryways
Rectilinear and Polygonal

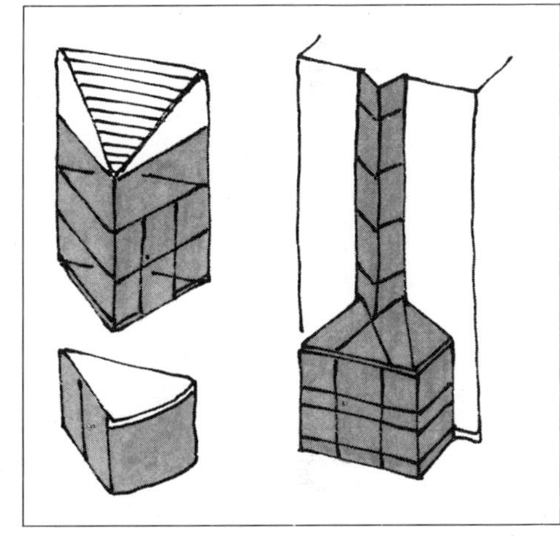

149

Triangular

Triangular forms show clearly the reasons for their design.

The lighting of the stairwell in the example at left is oriel-like and connects with the triangular, wind-cheating entryway. The inward turned triangular oriel at right is based on the same architectural motives, this time applied to the renovation of an old building. It is a practical and impressive construction, standing in sharp contrast to the spare white walls of the building.

Round entryways typically contain revolving doors.

Only one of the examples shown here has sliding doors. The continued advance of technology in the area of sliding glass doors allows increasingly practical applications of automatically opening doors.

It is relatively easy to construct round forms if small vertical sections are used.

It is especially easy to face these spaces internally with wood paneling or slatting, although the substructure is still problematical.

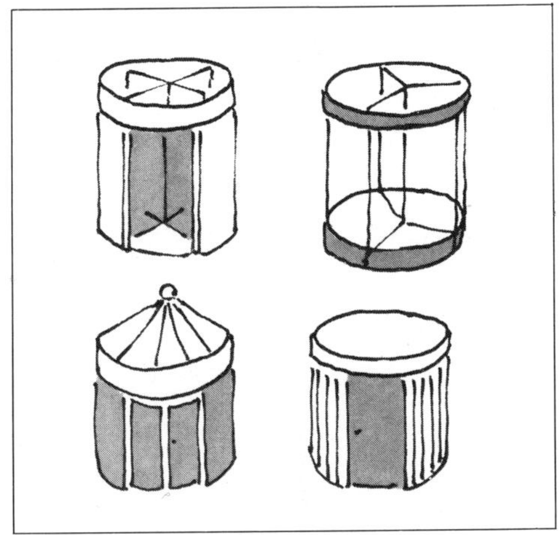

151

Round

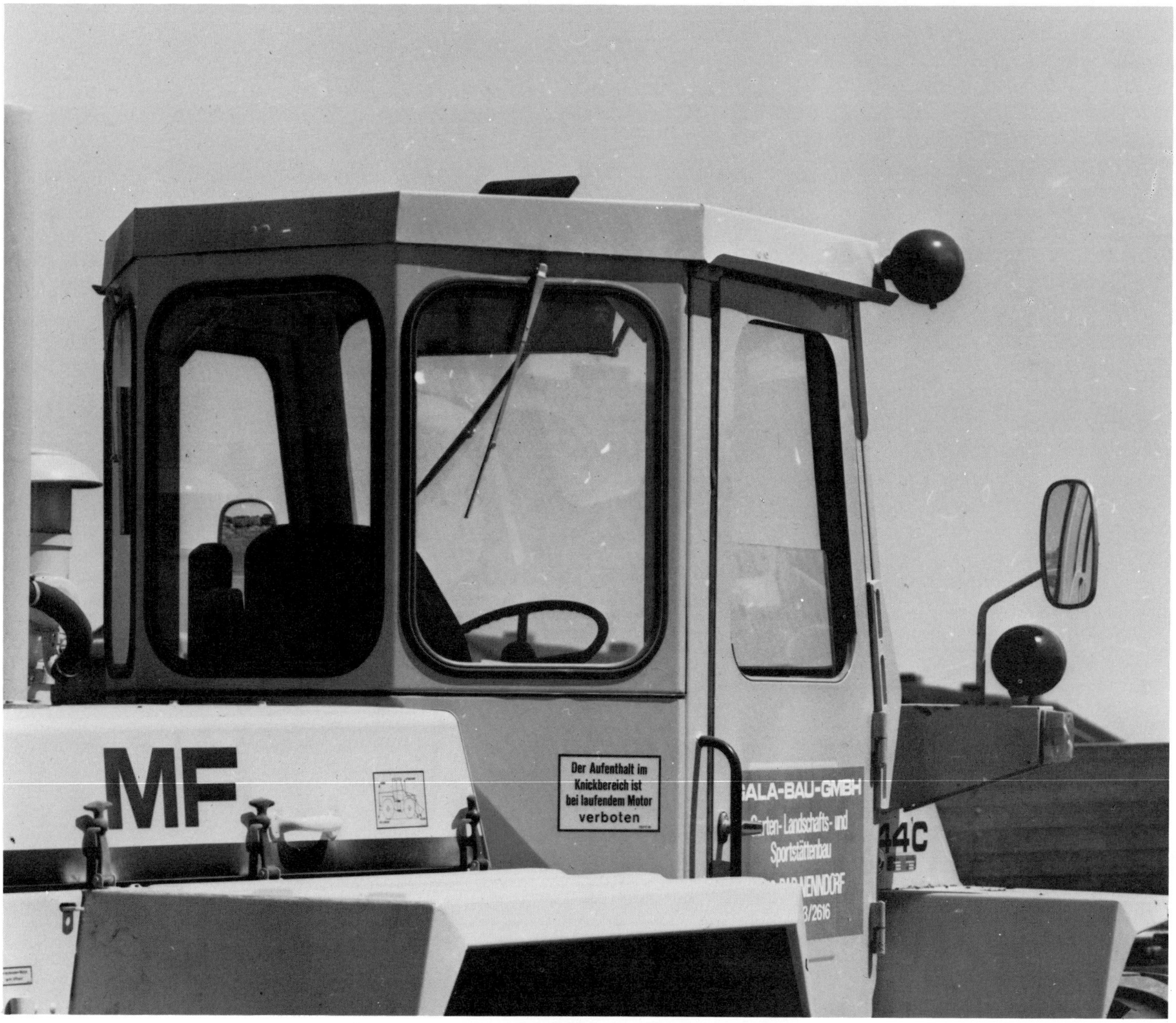
MF
Der Aufenthalt im Knickbereich ist bei laufendem Motor verboten
ALA-BAU-GMBH
Sportstättenbau
44C

60 85 99-29 320-6
Db

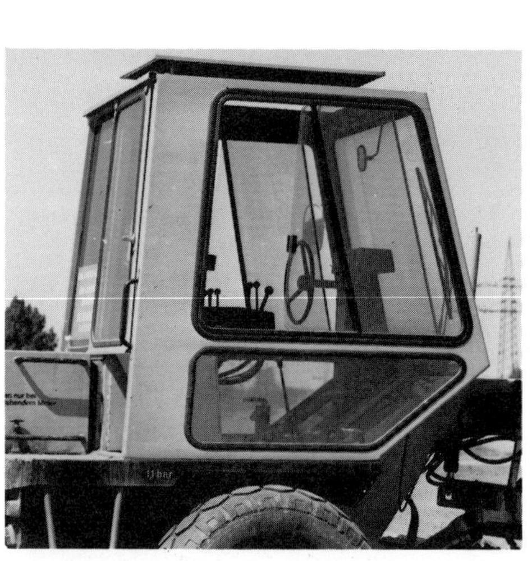

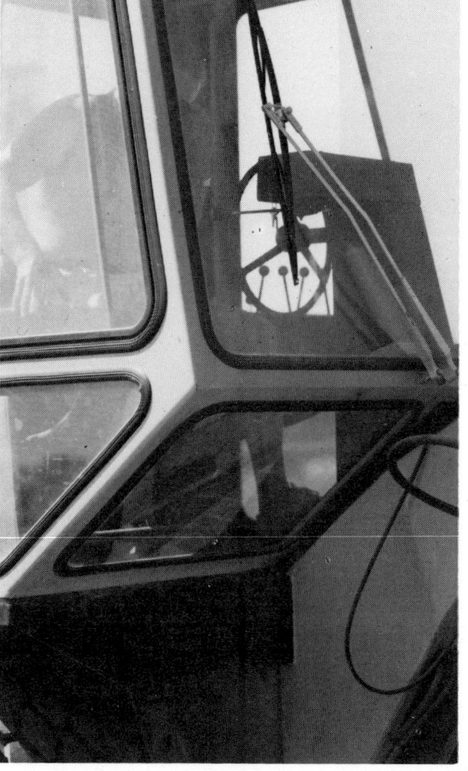

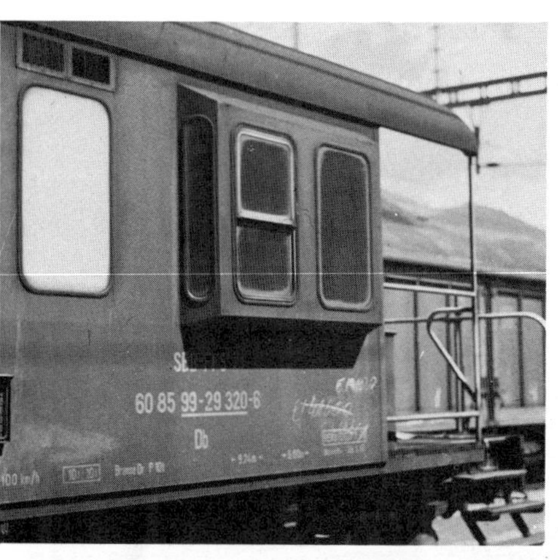

If we consider all the examples of oriels, dormers, and facade extensions that we have seen thus far, the windshields on the gondolas of ski lifts and the cockpits of planes and racing vehicles must strike us as analogous.

Because oriels and the protective windshields of a variety of objects serve the same purpose—increasing the occupant's range of vision in several directions—their forms are closely related. Their implementation differs, however, and therein lies their relevance to the subject of this book.

In the construction of a chassis, we are used to adding complicated individual parts in a rational order.

In construction, builders must use similar methods in order to construct facade extensions at less cost. On the international scene, it is already possible to buy prefabricated "clip-on" oriels.

153

Windshields

CONSTRUCTION EXAMPLES: WOOD, STEEL, AND ALUMINUM

Oriel constructions can be realized in wood as well as metal. Wood framing is relatively strong and al-lows impressive visibility. Its insulating qualities are excellent.

Steel framing can be very unobtrusive, allowing constructions with a very ethereal appearance. The cold-conducting factors of metal frames are so minimal, even when compared with thermal glass surfaces, that no special allowance need be made for them.

Aluminum framing comes both insulated and non-insulated. Although somewhat difficult to work with, it does provide an elegant and precise contour element. Veneering materials can be mounted through the use of beveled aluminum moldings.

Steel framing makes even complicated oriel constructions easy to

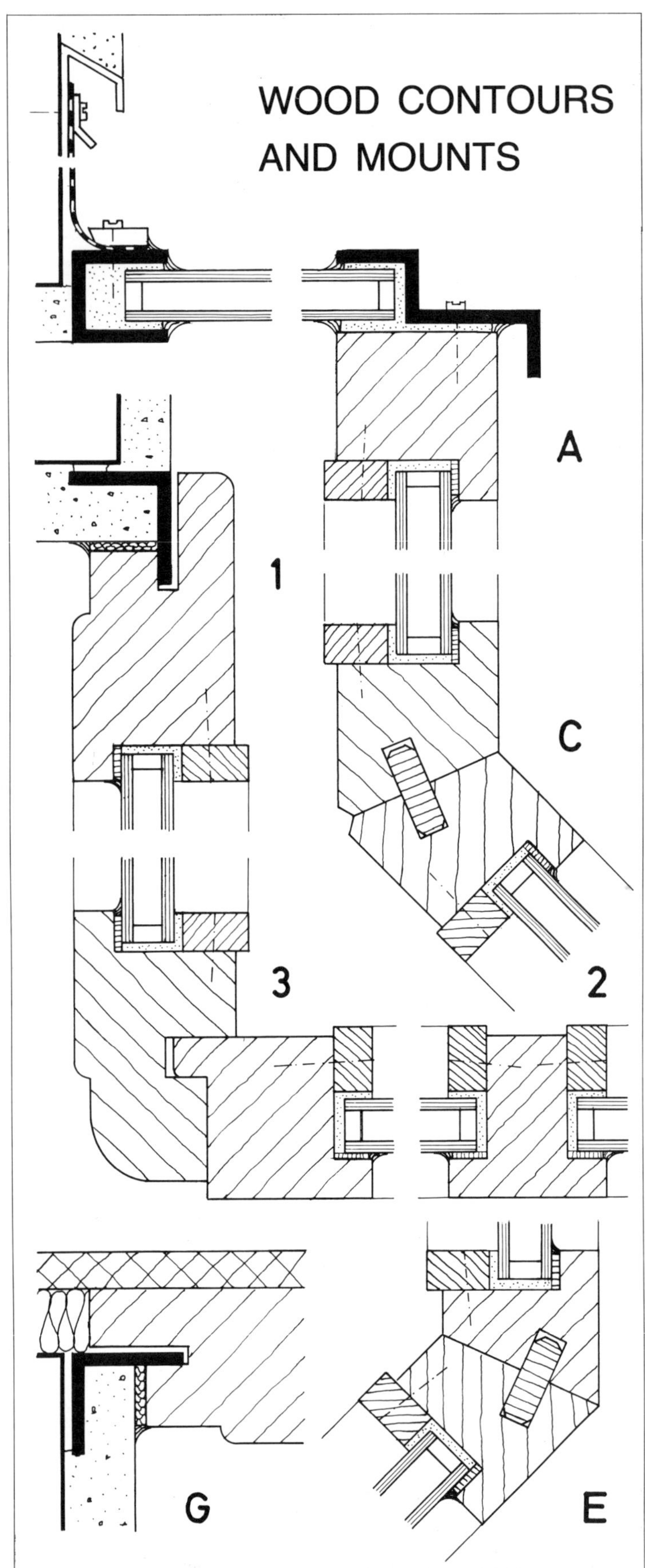

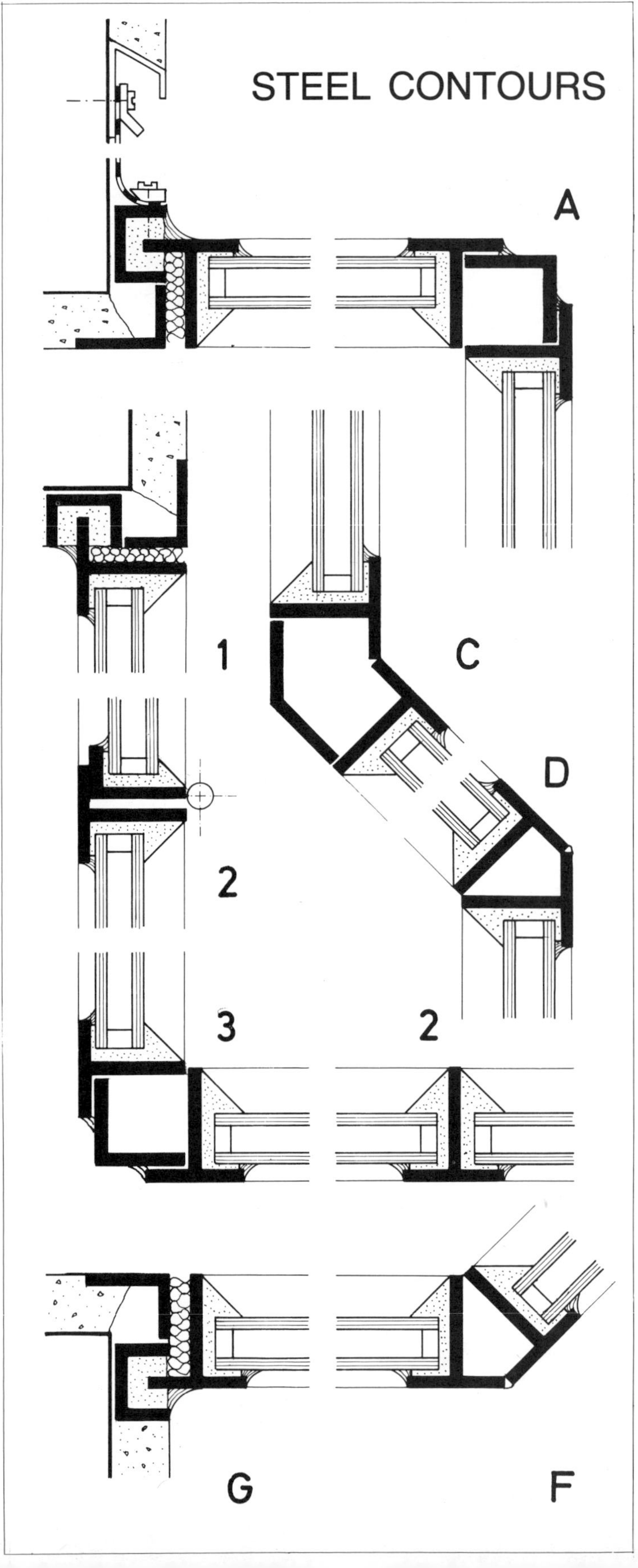

realize. Rough joinings can easily be smoothed, filled, and coated, and will provide durable joints.

Aluminum framing, to the contrary, demands exact measurements and is very difficult to rework.

Wood framing can be done with all the standard woods, such as fir and pine. The wood is usually lacquered or sealed by painting.

Combination constructions can be practical and economical. The finishing work must be carefully undertaken, however, to avoid leaving areas where water might leak or collect, leading to rot and mold.

Wood can be treated to prevent fading or bleaching, and steel should be galvanized. Aluminum is usually anodized and, more re-

cently, coated with a stove lacquer.

Coloration is no longer limited by the choice of materials. Glazings, however, must be carefully sealed or caulked.

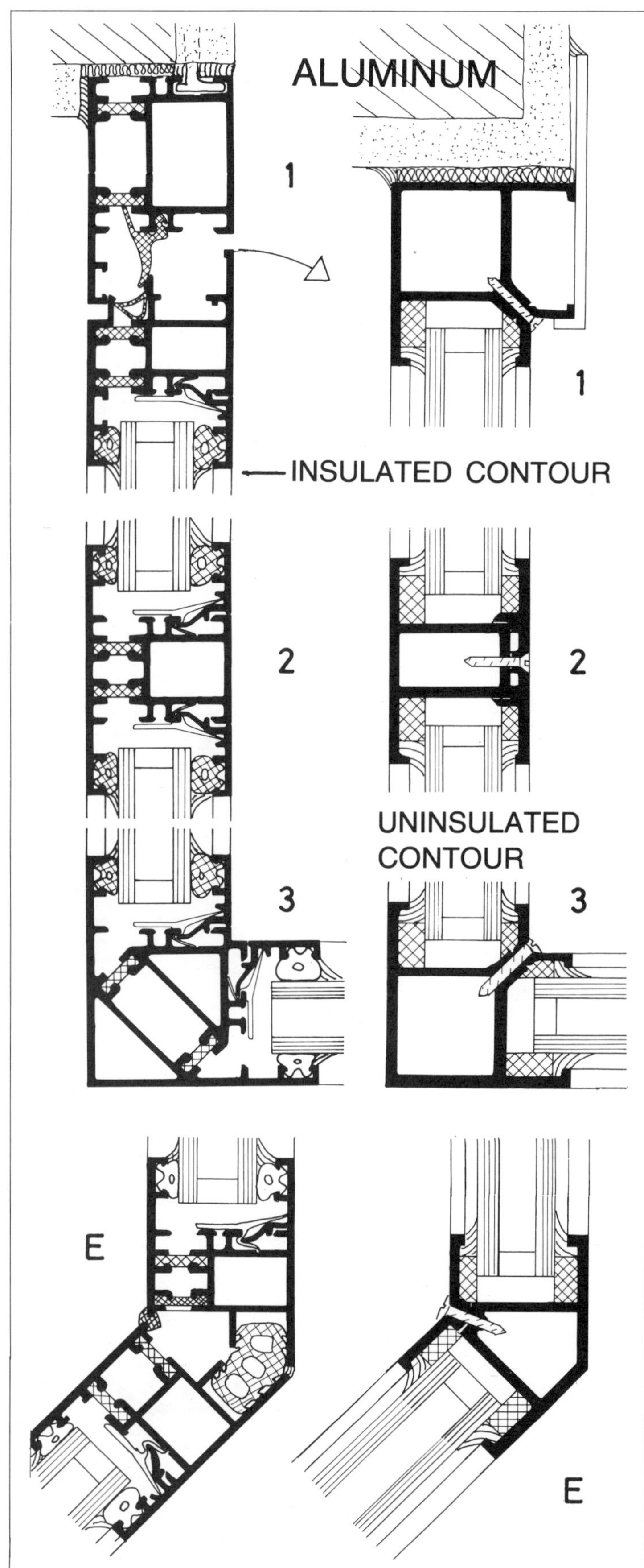

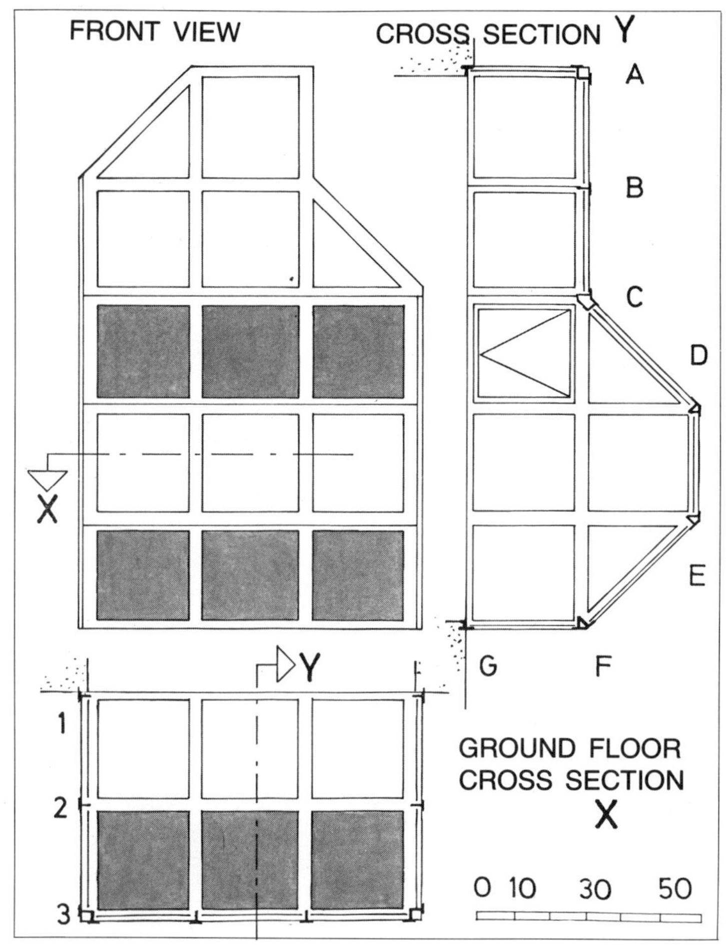

155

PHOTO CREDITS

Bennet & Powell, London (3)

Dieter Blase, Hamburg (281)

Bofinger, Brunswick (7)

Brecht-Einzig, Wimbledon (3)

Caseriego, Madrid (3)

False Creek, Vancouver (5)

Nils Friese, Aachen (4)

Ursala Dannien, Lübeck (1)

Gerhard Diel (3)

John Donat, London (3)

Dyckerhoff Cement Works (1)

Bert Federau, Hamburg (5)

Fischer-Daber, Hamburg (1)

Graaf & Schweger, Hamburg (4)

Inge Goertz-Bauer, Düsseldorf (1)

Robert Häusser, Mannheim (3)

Hölscher, Hanover (2)

Krawina, Vienna (1)

Lill, Hanover (4)

Moore, New Haven (2)

Mero, Würzburg (1)

Sigrid Neubert, Munich (1)

Planen & Bauen, Hamburg (8)

Thomas Pracht, Mergoscia (172)

Röhm GmbH Chemical Works (9)

Elke Rodemeier, Cologne (11)

Wilfried Rabanus, Munich (1)

Jürgen Röpunus, Munich (5)

Rübartsch, Heidelberg (1)

Heike Seefeld, Hanover (1)

City Construction Bureau, Hanover (2)

Steel Advisory Bureau, Düsseldorf (2)

Friedhelm Thomas, Meerbusch (3)

Administration Bureau, Kiel (2)

Heinrich Wolf, Coblenz (1)

PICTURE CREDITS

Cover: Apartment building in Rotenburg on the Wümme River. Architect: Planen & Bauen.

16 Historical half-timbered oriel in Hanover-Münden, Höxter Ahfeld, and Bevern (l.).

18 Rectangular oriel in Hanover, Wetterau, Göttingen, Cologne, and Herdecke (r.).

20 Rental units, Lucerne (l.).
Apartment houses, Hanover (r.).

22 Half-round oriels
Old people's home and department store, Hanover (u.l.).
Historical rental units, Hanover (l.l.).
Ihme Center and terrace houses, Hanover (u.r.).

24 Tilted oriels
City hall of Bensberg. Architect: Gottfried Böhm (u.l. and u.r.).
SOS Children's Village, apartments in Esslingen and Stuttgart (l.l.).
Public swimming pool in Springe (l.r.).
School in Hanover.

26 Apartment house in Cologne. Architect: Erich Schneider-Wessling.

28 Triangular oriels with supports
Rental units in Düsseldorf. Architect: Wolfgang Döring (l.).
Commercial building in Bremen. Architect: Meyer-Burg (u.r.).
Department store in Offenburg (l.r.).

30 Triangular oriels
Calwer Passage in Stuttgart. Architect: Kammerer & Belz (u.l.).

Savings bank in Staufen (l.l.).
City hall in Herdecke. Examples from Wetterau. Architect: Gehse & Detlef (u.r.).
Apartment house in Stuttgart. Architect: Wolz (l.r.).

32 Commercial building in Esslingen. Architect: Jörg Könekamp (u.l.).
University of Berlin. Architect: Daniel Gogel (m.l.).
Conservatory of Music, Hanover (u.r.).
South Shore Apartment Complex, Planer False Creek, Vancouver (l.r.).

34 Historical brickwork oriel
Examples from Hanover, including the old city hall (r.).

36 Skylights
Bank in Osnabrück. Architect: Haffkemeyer & Fangmeyer (u.r.).
University of Bremen. Architect: Me-di-um (l.l.).
Community hall in Lübeck. Architect: Dannien & Fendrich (u.r.).
Kindergarten in Berlin.
Rental units in Locarno (l.r.).

38 City construction project on Gross St. Martin in Cologne. Architect: J. & A. Schürmann (l.).
Apartment building in Stuttgart. Architect: Meyer (l.l.).
Conservatory passage in Göttingen. Architect: Bofinger (l.r.).
Apartment building in Osnabrück (l.r.).
Oriel in Vancouver.

40 Terrace oriels
Apartment house in England (l.).
Professional building in Arzen (l.l.).
Garden house in Hamburg. Architect: Gerhart Laage (u.r.).
Senior citizens' center in Grebenstein. Architect: Jourdan & Müller (l.r.).

42 Console-like oriels
Rental units in Cologne (u.l.).
Conservatory passage, Göttingen.

Commercial building in Hanover.
Calwer Passage in Stuttgart. Architect: Kammerer & Belz (l.r.).
City building project, Vancouver. (l.r.).

44 School in Hanover-Mühlenberg (u.).
Commercial building in Hanover (l.l.).
Penthouse in Göttingen.
Ecumenical center in Stuttgart-Neugereut (l.r.).

46 Off-set oriels
Community complex, Planer False Creek, Vancouver (l.l.).
Museum in Paderborn. Architect: Gottfried Böhm (u.r.).
School in Liechtenstein (l.r.).

48 Oriels in the eaves
Rental units in Hamburg. Architects: Planen & Bauen (l.).
University of Bremen. Architect: Assmann & Störner (l.r.).

50 Municipal Theater, Wiesbaden. Architect: H.W. Hämer (l.).
Kaufhaus in Freiburg. Architect: Heinz Mohl (r.).

52 Apartment house in Lemgo. Architect: Donat (l.).
Renovations in the old part of Lemgo. Architect: Walter von Lom (r.).
New construction area in Hamburg (l.r.).

54 Wraparound oriels
Apartment house in Rotenburg on the Wümme River. Architect: Planen & Bauen.
Conservatory passageway on Kempten (r.).

56 Historical sandstone oriels
Castle and Laveshaus in Hanover (u.l.).
City halls of Paderborn and Bremen (l.l.).
City hall of Lemgo (l.r.).

58 Interior oriels
Schoolyard in Hanover-Laatzen (l.).
Vacation houses, Sea Ranch, California. Architect: Charles Moore (r.).

INDEX